Natural theory of relativity, inertia, gravitation & gravitivity

On the Shoulders
of
Tahir

Copyright © Fayaz Tahir 2023
First published 2016
https://www.youtube.com/@Gravitation_Gravitivity

All praises and thanks be to ALLAH Kareem - The Supreme - The Lord *of* Muhammed (pbuh), WHO inspired Tahir through Ilham and taught him with the pen which he knew not.

Contents

1. Introduction

The current theories of relativity are separated by an unpassable gulf which makes the understanding of relativity very difficult and confusing for ordinary readers from all trades of life. Moreover, the mathematics involved is complicated and beyond the reach of people unrelated to the physics profession. People from all walks of life are interested in learning about relativity but they just can't. This theory has been woven with the fabric of inertia and gravitation in such a manner that ordinary people who know what Alkashi's Theorem is and have a little bit introduction with the concept of derivative can easily understand the mysteries of the changing orbits of planets and the double deflection of a photon when it passes near a mass in an easy and elegant manner. After reading this theory no confusion is left in the minds of the readers. Moreover, the actual picture is also seen which remained hid from the sight of people for the last one hundred years. People like lawyers, schoolteachers, nurses, college teachers, university teachers, school kids, tax agents, property dealers, doctors, engineers, scientists, managers of different businesses, shop owners, business owners, scholars of all categories, book publishers, book binders etc. can easily understand my theory without leaving *any* ambiguity in their minds. Once they have understood my theory, they have understood the laws of creation of space and matter by the Lord of Muhammed (pbuh) – ALLAH – Kareem – The Supreme. My theory is the greatest for the great and unwelcoming for the jealous and mean sons and daughters of Adam & Hawwa. **Any place on earth in the form of a house, library, school, college, building, university that doesn't house my book is full of POVERTY in wisdom and knowledge of Mathematics, Physics and Natural Philosophy.** My theory will strike the world like thunderbolts. People who can't afford heart palpitations should stay away from my theory.

The current theories on relativity have already frustrated and constantly disappointing generations of people from all walks of life without any slightest doubt and people have been forcefully made to believe that nothing better can pop up and it is impossible to produce something better which I have proven to be a lie. Instead of encouraging people to come up with something better than the current theories they have been discouraging them from believing in what has already been done. I wanted people to come out of this darkness and I was eventually victorious. In the derivation of the Relativistic Time Dilation formula a totally new series expansion of functions has also been introduced in the Calculus of Ibn-ul-Haishem. Moreover, the primordial centrifugal/centripetal accelerations and their associated vectorial and legalized (imaginary turned real) values of velocities have also been *eventually* found after the demise of Islamic Classical Mechanics when Sir Ibne-Bajjaj passed away as below:

$$\frac{V^2\cos^2\beta}{r} - \frac{d^2r}{dT^2}$$

&

$$i^2\left[\frac{V^2\cos^2\beta}{r} - \frac{d^2r}{dT^2}\right]$$

$$V' = -(V\sin\beta + V\cos\beta) \text{ (Tahirian Centripetal Velocity)}$$
$$\frac{d}{dT}[-m(V\sin\beta + V\cos\beta)] = \kappa G \frac{Mm}{r^2}$$

$$V'' = V\sin\beta + V\cos\beta \text{ (Tahirian Centrifugal Velocity)}$$

$$\text{Abul Barkat Al-Baghdadi} = \text{Abū al-Fath Abd Al-Rahman Mansūr Al-Khāzini} + \text{Tahir} + \text{noton} + \text{anstan}$$

Where:

$$i = \sqrt{-1} \equiv \text{Rotation of } 90°$$
$$i^2 = -1 \equiv \text{Rotation of } 180°$$

The applied force is directly proportional to the rate of change of momentum and acts in the direction in which the change acts.

I always believed that it is just a matter of finding the velocities associated with Tahirian Centripetal/Centrifugal Forces and thus I have been eventually able to find them once and forever! Invincible and indisputable proofs of the legalized values will follow in the later formulations of the theory. The legalized System of Values of Islamic Mathematics is so important in the understanding of Islamic Physics that I had to dedicate a splendid article for that in my Revolutionary Works on Natural Philosophy.

If the above is not true algebraically then the entire Classical Mechanics of Abul Barkat Al-Baghdadi and the entire Gravitation of Abū al-Fath Abd al-Rahman Mansūr al-Khāzini is dismantled! The entire inertia of Ibn Sina is dismantled! The entire Calculus of Ibn Haishem is dismantled! The entire mathematics of AlKhawarizmi is dismantled! The entire Revolutionary Works of Tahir are dismantled! Tahir Angle (TA) β and Tahirian Rotations (TR) should have been incorporated by the Naives (noton, anstan, maglel, copernicooni, kapil, taka bro, hoking etc.) in their commentary of Classical Islamic Physics which they couldn't! What a remarkable series of Naives!

Can anyone produce equations more elegant and delicate than the above?

My book has dwarfed the commentary of the Naives!

I believe that they were created by ALLAH Kareem - the Supreme, to be destined for matter, even before smoke, dust or matter itself were created. These are the accelerations of inertia. Above all, the Nature has opened his heart to Tahir's Pen in the form of the mother of all equations as below:

Tahir Field Equations *of* all Motion

$$\frac{dv}{dT} = [\frac{d^2r}{dT^2} - r[\frac{d\theta}{dT}]^2]\sin\beta + \overline{[r\frac{d^2\theta}{dT^2} + \frac{dr}{dT}\frac{d\theta}{dT} + \frac{dr}{dT}\frac{d\theta}{dT}]}\cos\beta$$

$$\frac{d\beta}{dT} = \frac{1}{v}[\frac{d^2r}{dT^2}]\cos\beta - \frac{1}{v}[r\frac{d^2\theta}{dT^2} + \frac{dr}{dT}\frac{d\theta}{dT}]\sin\beta$$

The above equations are the Tahirian Equations of all Motion or the General Field Equations. Overbar doesn't mean mean or average value. It means the terms to be taken collectively for some reason that will be made crystal clear. The actual story was the Story of Inertia of Ibne Sina, the couple of inertial accelerations of Abul Barakat Al-Baghdadi and the Inertial action and reaction principle of Ibne- Bajjaj and the Great Derivations of Tahir and *not* of the significant force of attraction between (two or more) particles as commented by the northerners (usa, russia etc.) on Western and Eastern Classical Islamic Physics.

Postulates of Natural Relativity & Gravitivity

1. The laws of physics are the same in all inertial frames.
2. No two inertial frames are identical, indistinguishable or equivalent unless or until they move at the same inertial velocity. It is just another mathematical way of writing the same frame and hence it reduces to the same one as before and does not produce another mathematically distinguishable frame (See page 67).
3. The speed of light or gravity is the same when it is measured and recorded by spacetime clocks at rest (hypothetical) or coordinate clocks at rest or in motion with the photon (hypothetical) *with infinite speed of information*. Coordinate clocks may even run with the photon because the photon itself has a coordinate or spacetime clock i.e. no proper time, running with it after it has colossally accelerated to c in the tiniest fraction of a second.
4. The speed of light or gravity is measured differently in different inertial frames with coordinate or spacetime clocks running with the object (an object in motion constitutes a frame of reference) at $V_1, V_2, V_3, \ldots V_n <$ C or proper time clocks running with the object at V_1, $V_2, V_3, \ldots V_n <$ C measuring *with finite speed of information i.e. the speed of light itself.* (See page 67)
5. The values of the coordinate velocities $V_1, V_2, V_3, \ldots$ V_n are measured with an infinite speed of information by coordinate clocks and hence set the stage for my theory.

Speed of gravitational wave S = 299 792 458 m/s (Gravity)
Speed of light photon / wave C = 299 792 458 m/s (Relativity)

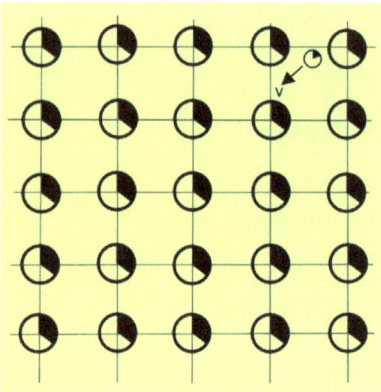

***Figure 0**. Coordinate clocks, spacetime clocks, fixed clocks or hypothetical clocks. Also shown is an inertial clock ticking slower in motion.*

Law of Binary Attractions

The Sun and the Planet i.e., two particles, attract each other with a force directly proportional to the product of the product of their masses ($m_1 m_2$) and the Tahir Gravitivistic Constant (κ) and inversely proportional to the square of their distance apart (r).

$$F = \left(1 + 3\frac{v^2}{s^2}\right)\frac{Gm_1m_2}{r^2}$$

[Exact/close form of the Tahirian Gravitivistic Force]

where $v = v_1 + v_2$ & $r = r_1 + r_2$

Further Laws of Tahirian Binary Stars Cosmology are inside the book and complete the highly misunderstood phenomenon of Binary Attractions *uselessly* debated by many.

$$\frac{v_{12}^2}{2} - \frac{v_{11}^2}{2} \sim GM(\frac{1}{r_{12}} - \frac{1}{r_{12}})$$ Static orbits

$$v_{22}^2 = \frac{s^2}{3t_2^2}\{(1 + 3\frac{v_{21}^2}{s^2}t_2^2)e^{[\frac{6Gm_1}{s^2}(\frac{1}{r_{22}} - \frac{1}{r_{21}})]} - 1\}$$ Tahirian orbits

$$v_{21} = \frac{s}{t_2}\sqrt{\frac{1}{3}(e^{\frac{6Gm_1}{s^2r_{11}}} - 1)}$$ Tahirian Escape Speed

$$6\pi\frac{G^2m_1^2}{t_2^2s^2h_2^2}$$ Tahirian Precession in radians per revolution

Where t is the Tahir Correction/Constant of Binary Star Systems in Tahirian Cosmology

$$t_2 = 1 + \frac{m_2}{m_1}$$

where $m = m_1 + m_2$

Remarkable derivations will follow. The precession of the planets has been found along with the double deflection of photon/wave passing near the sun in an excellent manner namely stationary sun and binary star motion *both*. My theory explains the precession of Pluto and its partner where general relativity fails as per Wikipedia because of the addition of Tahir Constant of Binary Star Systems as above in the gravitivistic precession formula. Moreover, the Janum Kundli of relativity and Gravitivity has also been constructed as below. *The precession of Mercury due to the tugs of other planets has also been remarkably found by the Tahirian integral. The method employed is remarkable and not present, anywhere else, on the back of the earth. Moreover, it is accurate to the extent of Tahironian Gravitivity!*

Standard Model of Relativity and Gravitivity

ALLAH Kareem – The SUPREME showered his utmost mercy on his Highness Sir Muhammed Ghiyasuddin Jamshed Al-kashi by giving him the greatest of all the theorems in mathematics and physics and natural philosophy i.e., Theorem de Al-Kashi. The macroscopic mysteries of the Cosmos could never ever be understood completely by Tahir if Tahir had not utilized the 'Gem of Jamshed.' His Jem can also be called "The First Complete Metric" [1].

If symmetry had been the cause of creation, then there would not have been the need for maintaining a balance be-

Table 1. Standard Model of Gravitivity and Relativity.

cause balance is requirement of *asymmetry*! We have been

Janam Kundli of Gravitivity & Relativity	Under Speed of Information (Gravitivity) and Speed of Photon (Relativity) s = c= 299 792 458 m/s (c in vacuum only)	Under Speed of Gravitational information (Gravitivity) s = 299 792 458 m/s everywhere
T	Coordinate Time Dilation of Photon particle or any other hypothetical coordinate particle due to measuring the information with the speed of information/gravity. $$T = \frac{t}{\left(1 - \frac{v^2}{c^2}\right)^2}$$	Coordinate Time Dilation of ordinary particle due to measuring the information with the speed of information. $$T = \frac{t}{1 - 2\frac{v^2}{s^2}}$$
T′	No Proper Time Dilation for Photon particle or any other hypothetical coordinate particle. Proper Time is sandwiched between the two Coordinate Times t and T.	Proper Time Dilation due to the speed of information $$T' = T\left(1 - \frac{v^2}{s^2}\right)$$

created unsymmetrically. Rather there is a sublime mix of symmetry and asymmetry in our creation. If symmetry had been the intention of ALLAH Kareem – the SUPREME, then HE would have placed the heart in the centre of the body and, also, the sun at the centre of the elliptical path of the planet mathematically and not at one focus (which is wrong – I have proven it mathematically in Article 4) and would have shown us other directions of calculations. Where mathematics produces symmetry, it simply means it is the wrong direction of calculations or pseudo calculations. This asymmetry has been

intentionally put in our cause of creation so that we may show our capability to keep a balance in life. We are constantly being tested by our Lord, our Rub –ALLAH Kareem the SUPREME. HE wants to see who does the good deeds and sends them to HIM before he/she comes to HIM by keeping a symmetric balance in life. HE has created us unsymmetrically but wants us to behave symmetrically by keeping a balance in life. Why because HE is the Master of everything, and HE does what He pleases. Remarkable equations and results will follow regarding the above equations in Miracle Article 3.

It took me eleven years to answer my questions and convince myself because I was busy with other things as well. Relativity is the scientific study of the speed of light and gravitivity is the scientific study of the speed of gravity. It is amazing because both the speeds are the same.

The velocity of the inertial particle while in motion at constant velocity is measured with respect to two clocks namely the coordinate time clock but in motion and NOT ticking slower i.e. an imaginary clock or stationary coordinate clock or spacetime clock, and the proper time clock which is ticking slower while in motion. This must have been considered by the alien while deriving the work done in inertia. But on the contrary, he postulated that the velocities are basically the same or indistinguishable and allotted the time dilation factor to mass instead – hence the virus of relativistic mass and hence declaring that there is no absolute rest. Ridiculous, isn't it? I have found the equation of absolute rest and is given in Miracle Article 3.

I believe that the speed of waves in magnetoelectric theory is the speed of gravitational information. The speed of light in vacuum has the same value as that of the speed of gravitational waves. The speed of gravity doesn't change anywhere but the speed of light does change. I believe that light has a particle of mass, if the data taken from the web is correct in Article 7.4, and according to my theory it is given below whereas there is no particle associated with gravitational information speed. It is purely the speed of a wave sensation.

Mass of photon (in light) $\sim 1.295 \times 10^{-32}$ Kg

My Theory is extremely simple to understand even by a post-doctoral toddler who knows simple calculus from school. At the same time, its results are simpler and more straight - forward than that of other theories. My book is not a compilation of the thoughts of other people. Over 99% of formulation given herein in Kitab-e-Tahir belongs to Tahir. My book is in perfect agreement with the Fundamental Principles of Motion by Ibn-e-Sina, the Father of Inertia and Gravitation.

I have coined the term gravitivity which is defined as the scientific study of the speed of gravity or gravitational wave. Gravitivity and Inertia are the two wings of gravitation. The Tahirian Gravitational Model, as provided by nature, is fundamentally incomplete without its wings. As a matter of fact, these wings are related to each other very closely. Both produce a beautiful picture of Natural philosophy.

All five parameters of projectile motion have been conquered – namely the time of flight, the maximum radial distance it travels from the Centre of the Earth or the gravitating mass before returning, equation of the trajectory, the optimal angle of launch to produce maximum range and the maximum range. All motion under gravity is projectile motion whether it is Tahir's Projected Apple or the orbiting moon. Moreover, the coordinate time under gravitation and not just under inertia has also been found by solving the Tahirian Differential Equation.

---

1. Gravitivistic times of flight of a projectile

$$T = 2 \left| \left[\frac{\sqrt{R}}{cu} + \frac{b}{2c\sqrt{-c}} \arcsin \frac{2c+bu}{u\sqrt{b^2-4ac}} \right] \Big|_{\frac{1}{R}}^{\frac{1}{r_0}} \right| \quad \{c<0, \ \Delta<0\}$$

$$T = 2 \left| \left[\frac{\sqrt{R}}{cu} + \frac{b}{2c\sqrt{-c}} \arctan \frac{2c+bu}{2\sqrt{-c}\sqrt{R}} \right] \Big|_{\frac{1}{R}}^{\frac{1}{r_0}} \right| \quad \{c<0\}$$

$$T' = 2 \left| \left(1 - \frac{v^2}{c^2}\right) \left[\frac{\sqrt{R}}{cu} + \frac{b}{2c\sqrt{-c}} \arcsin \frac{2c+bu}{u\sqrt{b^2-4ac}} \right] \Big|_{\frac{1}{R}}^{\frac{1}{r_0}} \right| \quad \{c<0, \ \Delta<0\}$$

(Time shown by the clock travelling along the projectile and ticking like an inertial clock)

$$T' = 2 \left| \left(1 - \frac{v^2}{c^2}\right) \left[\frac{\sqrt{R}}{cu} + \frac{b}{2c\sqrt{-c}} \arctan \frac{2c+bu}{2\sqrt{-c}\sqrt{R}} \right] \Big|_{\frac{1}{R}}^{\frac{1}{r_0}} \right| \quad \{c<0\}$$

(Time shown by the clock travelling along the projectile and ticking like an inertial clock)

$$t = 2 \left| \left(1 - 2\frac{v^2}{c^2}\right) \left[\frac{\sqrt{R}}{cu} + \frac{b}{2c\sqrt{-c}} \arcsin \frac{2c+bu}{u\sqrt{b^2-4ac}} \right] \Big|_{\frac{1}{R}}^{\frac{1}{r_0}} \right| \quad \{c<0, \ \Delta<0\}$$

$$t = 2 \left| \left(1 - 2\frac{v^2}{c^2}\right) \left[\frac{\sqrt{R}}{cu} + \frac{b}{2c\sqrt{-c}} \arctan \frac{2c+bu}{2\sqrt{-c}\sqrt{R}} \right] \Big|_{\frac{1}{R}}^{\frac{1}{r_0}} \right| \quad \{c<0\}$$

Where:

$$a = \frac{6G^2m_1^2}{t_2^2 s^2} - h_2^2$$

$$b = \frac{2Gm_1}{t_2^2} - \frac{12G^2m_1^2}{t_2^2 s^2 r_{21}} + \frac{6Gm_1 v_{21}^2}{s^2}$$

$$c = v_{21}^2 - \frac{2Gm_1}{t_2^2 r_{21}} + \frac{6G^2m_1^2}{t_2^2 s^2 r_{21}^2} - \frac{6Gm_1 v_{21}^2}{s^2 r_{21}}$$

&

$$R = au^2 + bu + c, \quad \Delta = 4ac - b^2 \ \& \quad u = \frac{1}{r}$$

2. Gravitivistic maximum radial distance travelled by the projectile

$$r_0 = \frac{B + \sqrt{B^2 + Ah^2}}{-A}$$

Where:

$$A = v_{2R}^2 - \frac{2Gm_1}{t_2^2 r_{2R}} - \frac{6Gm_1 v_{21}^2}{s^2 r_{2R}} + \frac{6G^2m_1^2}{t_2^2 s^2 r_{2R}^2}$$

$$B = \frac{Gm_1}{t_2^2} - \frac{6G^2m_1^2}{t_2^2 s^2 r_{2R}} + \frac{3Gm_1 v_{21}^2}{s^2}$$

$$C = \frac{6G^2m_1^2}{t_2^2 s^2} - h_2^2$$

$$h = \textbf{check inside book}$$
$$v_{2R} = v_{21}$$

3. Gravitivistic equation of the trajectory of a projectile

$$r_{22} \approx \frac{1}{\dfrac{-b + \sqrt{b^2 - 4ac} \sin[\sqrt{p}\,\theta_{22}]}{2a}}$$

$$p = 1 - \frac{6G^2m_1^2}{t_2^2 h_2^2 s^2}$$

Precessing trajectories/orbits

Where:

$$a = \frac{6G^2m_1^2}{t_2^2 s^2} - h_2^2$$

$$b = \frac{2Gm_1}{t_2^2} - \frac{12G^2m_1^2}{t_2^2 s^2 r_{21}} + \frac{6Gm_1 v_{21}^2}{s^2}$$

$$c = v_{21}^2 - \frac{2Gm_1}{t_2^2 r_{21}} + \frac{6G^2m_1^2}{t_2^2 s^2 r_{21}^2} - \frac{6Gm_1 v_{21}^2}{s^2 r_{21}}$$

$$t_2 = 1 + \frac{m_2}{m_1}$$

4. Gravitivistic value of the optimal angle of launch

$$\cos\beta_R = \frac{1}{R v_R} \sqrt{\frac{(2B + AR)}{-Ap} B}$$

Where:

$$A = v_{2R}^2 - \frac{2Gm_1}{t_2^2 r_{2R}} - \frac{6Gm_1 v_{21}^2}{s^2 r_{2R}} + \frac{6G^2m_1^2}{t_2^2 s^2 r_{2R}^2}$$

$$B = \frac{Gm_1}{t_2^2} - \frac{6G^2m_1^2}{t_2^2 s^2 r_{2R}} + \frac{3Gm_1 v_{21}^2}{s^2}$$

$$C = \frac{6G^2m_1^2}{t_2^2 s^2} - h_2^2$$

$$p = 1 - \frac{6G^2m_1^2}{t_2^2 h_2^2 s^2}$$

$$t_2 = 1 + \frac{m_2}{m_1}$$

5. Gravitivistic value of the maximum range of projectile

$$S = R\theta_0 = \pi R - 2R \frac{h}{\sqrt{-a}} \arcsin \frac{\left(\frac{2a}{R} + b\right)}{\sqrt{b^2 - 4ac}}$$

Where:

$$t_2 = 1 + \frac{m_2}{m_1}$$

$$a = \frac{6G^2 m_1^2}{t_2^2 s^2} - h_2^2$$

$$b = \frac{2Gm_1}{t_2^2} - \frac{12G^2 m_1^2}{t_2^2 s^2 r_{21}} + \frac{6Gm_1 v_{21}^2}{s^2}$$

$$c = v_{21}^2 - \frac{2Gm_1}{t_2^2 r_{21}} + \frac{6G^2 m_1^2}{t_2^2 s^2 r_{21}^2} - \frac{6Gm_1 v_{21}^2}{s^2 r_{21}}$$

My book explains everything in just one single theory and is in perfect agreement with the superb inertial principals of the Muslim Fathers of inertia and gravitation namely Ibne-Sina, Abul- Barakat Al-Baghdadi and Ibne-Bajjaj – the owners of the three fundamental Islamic Laws of Motion.

Ibne-Sina-noton First Law of Motion

A body continues in its state of inertial rest $v = 0$ (which is actually not possible for any particle of matter except with ALLAH Kareem – The Supreme – with whom all things are possible) or in gravitational motion $v \neq 0 \,\&\, v \neq constant$ unless it is compelled by gravitational force to come out of inertial rest $v = 0$ or compelled by inertial state due to Tahir Hypothetical Inertial Catastrophe (THIC or THIS(Tahir Screening under binaries)) to come out of gravitational motion $v \neq 0 \,\&\, v \neq constant$ and move in a straight line with a constant velocity i.e., $v = constant$.

Ibne-Sina- Tahir Universal Law(s) of Motion

A body continues in its state of uniform motion in a straight line unless compelled by some external force to act otherwise.

A body continues in its state of uniform (constant angular speed) rotational/spinning motion about a fixed axis unless compelled by some internal/friction or external force/torque or gravity induced frictional force (due to action (gravity) and reaction (tension or compression) at the surface of contact) to act otherwise.

Or simply as below

A body continues in its state of uniform rotational motion about a fixed axis unless compelled by some external force/torque to act otherwise.

Or combining the two together as below

A body continues in its state of uniform motion in a straight line or/and uniform rotational/spinning motion about a fixed axis unless compelled by some external force to act otherwise.

Here I would like to narrate my dream that I had in Pakistan before coming to Australia in which I saw the Poet of the East sitting on a bench right next to the Philosopher *of* the East and the West in a Masjid. When I looked on my right, I saw him sitting and as a result I got a bit surprised that he is Muhammed Iqbal. As soon as I looked at him, he said to me, "Fayaz Sahb Azan dayo," as if he had been waiting for Salat, after which the dream finished. I happened to go to that Masjid and sit on that bench. When I sat on that bench there was nobody else in the premises of the Masjid except me. I felt very deep in my heart that something is going to happen soon, and my shoulders are going to hold up what no man held before. When I stood for prayer after giving the Azaan, the day seemed to have stood still with me for a moment and I felt somebody standing in the suf behind me till I finished my Salat.

The failure of the homo sapiens in understanding Ibne-Sina is the cause of the naive thoughts that spread like a contagious disease. It is the same force of gravitation that determines the parameters of the sun and the planet, earth and the moon, Tahir's Rising Apple (projectile) and the earth, the earth and the falling apple! Remarkable calculations will follow regarding finding the force of gravitation between the Sun and the photon by utilizing the dual properties of the photon-wave with a mass of photon in the direction of propagation of the wave producing double the gravitational value of the deflection of photon while passing near the sun!

In the development of this book the theorems of the Muslim Fathers of science and technology namely Muhammed Ghyiasuddin Jamshed Al-Kashi and Ibn-Muadh Al-Jayyani have been utilized along with the significant commentary of the aliens on the science of infinitesimals resulting in the fundamental "Tahir Principle of Instantaneity." My formulae are bound to enter every habitat in the world where inertia, gravitation, relativity and gravitivity are taught without nonsense! I am leaving no stone after me behind which a male or female can hide themselves and dare to say that they are not Tahironians!

In short, I have been able to remove the debt from the naïve shoulders of humanity that it owed to the understanding of natural philosophy. The third equation of gravitation has been eventually found that was the last Act in the drama of gravitation of two body motion. My book has put to shame the knowledge of the learnt of My Time, before and after. I hope mankind will prove their sensibleness, if they are left with some, by pledging their loyalty in the fullest to Tahir's Anatomy of Relativity, Inertia, Gravitation and Gravitivity.

I always wondered why there is a bifurcation between the physics community, one in favor of special relativity and the other against it. Then there came a moment in my life when I made up my mind to find the reason behind it. But just making up my mind was not the solution. I had to make attempts, and so I did. I started exploring different scenarios, one after the other, and after working continuously for three months or so I eventually hit the bull's eye. There was a moment on me when after making many apparently unsuccessful attempts I was going to give up the wild goose chase. I had the final equation in front of me but still no solution. After I took the limit, I at once got what I was after - the relativistic inertial time dilation formula and so I started my theory.

I have based my theory on the five postulates that have gone before. The speed of light is the speed of our information and constant for all observers i.e. inertial and gravitational and I also believe that the measurements taken in a frame of reference, of physical quantities of another frame of reference in relative motion, with the speed of our information as universal constant and requiring definite amount of time to travel between two relativistic frames, are dilated. The same analogy can be better applied to the speed of gravitational information which is the same as c. The contributions of two dilations have been brilliantly shown to have produced the *close form* of the gravitivistic form of the Gravitivistic Force in History once and forever.

I have built my theory modestly and have not plunged into big results at once. I have tried my best to give as many clarifications and derivations as possible and I have been successful.

I know my new interpretation of relativity will not be easy to swallow for the old relativity guys. But I can give surety that it does not hurt in the least because I am an old relativity guy, and I didn't have much difficulty in embracing the new interpretation of relativity.

My inertial thought experiment that will come next is in perfect agreement with the symmetry laid by Tahir Transformations.

2. Simple Thought Experiment –Relativistic Inertial Time Dilation

Let us run a simple thought experiment. To determine the inertial time dilation of a coordinate clock (at one point in space observed by an observer, having the same kind of coordinate clock, in a relative inertial frame moving with constant velocity v. I have put two clocks in the frame at rest relative to the moving frame, with a spatial distance x apart. We need to consider the *contradiction* that a moving clock is not ticking slower but is ticking like a spacetime or coordinate clock fixed in space-grid at every location of its coordinates. We will then see how an inertial clock ticking slower in inertial motion comes into calculations between the two coordinate clocks. We will then calculate the time the moving observer measures of the two clocks according to his measurability with a coordinate clock. The coordinate clocks are at rest basically but for understanding we have set them in motion and are not ticking slower like inertial clocks in motion. The two equations describing the measurement of time of two different but synchronized coordinate clocks in the rest frame will be scrutinized in the limiting case when the spatial distance x between the two coordinate clocks at rest tends to zero, hence making the two different times merge into one observation equal in magnitude. So, while the moving observer moves away from one coordinate clock, it is simultaneously moving closer to the second.

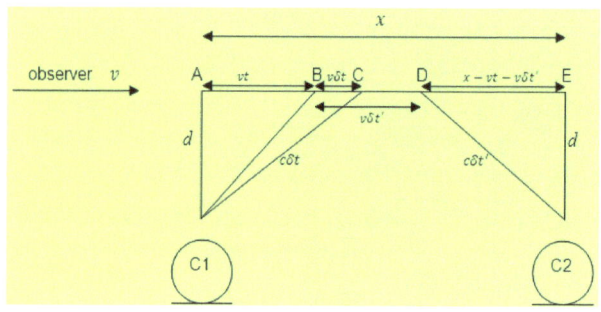

Figure 1. *Relativistic coordinate time dilation. Deriving the measured time interval T of a coordinate clock as a limit with the speed of our information in relativistic motion.*

In the above Figure 1 an observer is moving from the left to the right. Two coordinate clocks C1 & C2 synchronized with each other are placed x distance apart. When the observer approaches right on top of the clock at a height of d, the coordinate clocks C1 & C2 start synchronically. After an elapse of coordinate time t, the observer reaches point B in its moving frame. Both coordinate clocks at rest measure time t as both are synchronized with each other in the rest coordinate frame. Now, the observer tries to get the information of the time of C1 and by the time δt the information at the speed of light (hereafter considered to be the speed of our information) reaches him he has reached point C. Information speed takes definite amount of time (time light or light-like signals take to reach from C1 to observer at point C). The observer at point C records this time to be t_1 according to the equation below:

$$(vt + v\delta t)^2 + d^2 = (c\delta t)^2$$

$\delta t = t_1 - t$ & c is the speed of our information

Where δt is the definite amount of coordinate time in which the information reaches from C1 to observer at point C. In the same manner we see that when the same observer gets the information of the time of C2 he has reached point D. D does not necessarily need to be on the right of point C, it can be on the left of C depending upon the prevailing conditions between the observer, the coordinate clocks and the length of time interval being measured. The same observer records the time of C2 according to the same power of measurability with which he measured the time of C1. He measures it t_2 given by equation below:

$$(x - vt - v\delta t')^2 + d^2 = (c\delta t')^2$$

$$\delta t' = t_2 - t$$

Where $\delta t'$ is the definite amount of time in which the information reaches from C2 to observer at point D. Substituting the value of d^2 from one equation into the other, we get the following:

$$(x - vt_2)^2 + (c\delta t)^2 - (vt_1)^2 = (c\delta t')^2$$

$$x^2 - 2xvt_2 + (c^2 - v^2)(t_1^2 - t_2^2) - 2c^2t(t_1 - t_2) = 0 \quad (3)$$

$$\frac{x^2 - 2xvt_2}{t_1 - t_2}^2 + (c^2 - v^2)\frac{(t_1^2 - t_2^2)}{t_1 - t_2} - 2c^2t = 0$$

$$\lim_{\substack{x \to 0 \\ t_1 \to T \\ t_2 \to T}} \left\{ \frac{x^2 - 2xvt_2}{t_1 - t_2}^2 + (c^2 - v^2)\frac{(t_1^2 - t_2^2)}{t_1 - t_2} - 2c^2 t \right\} = 0$$

Let us now solve the above equation in the limiting case when x tends to zero. In doing so we will derive a small method by using Tahir Series Expansion of Al-Khawarizmic functions.

Now from the recently discovered Tahir Series Expansion of functions by Tahir at any arbitrary point (x,y) of two-variable functions we have:

$$f(x, y) = f(a, b) + \frac{\partial f(x, y)}{\partial x}(x - a) + \frac{\partial f(x, y)}{\partial y}(y - b)$$

$$- \frac{1}{2!}\frac{\partial^2 f(x, y)}{\partial x^2}(x - a)^2 - \frac{\partial^2 f(x, y)}{\partial x \partial y}(x - a)(y - b)$$

$$- \frac{1}{2!}\frac{\partial^2 f(x, y)}{\partial y^2}(y - b)^2 + \cdots$$

We can also write Tahir Series Expansion of two functions as below:

$$f(x, y) = f(a, b) + \left(hf_x + kf_y\right)\big|_{(x,y)}$$
$$- \frac{1}{2!}\left(h^2 f_{xx} + 2hkf_{xy} + k^2 f_{yy}\right)\big|_{(x,y)}$$
$$+ \frac{1}{3!}\Big[h^3 f_{xxx} + h^2 k(f_{yxx} + 2f_{xxy})$$
$$+ hk^2\left(f_{xyy} + 2f_{yyx}\right) + k^3 f_{yyy}\Big]\Big|_{(x,y)} - \cdots$$

$$f_{yxx} \neq f_{xxy}$$
$$f_{xyy} \neq f_{yyx}$$
Where $h = x - a$ & $k = y - b$

Here we have used the Tahir Convention and Tahir Fundamental Theorem of Partial differentiation. Both were the failures of naïve humanity in the understanding of Calculus.

Tahir Convention of writing higher order partial derivatives is:

$$\frac{\partial f}{\partial x} = f_x$$

or

$$\frac{\partial}{\partial y}\frac{\partial f}{\partial x} = \frac{\partial^2 f}{\partial y \partial x} = f_{yx}$$

or

$$\frac{\partial}{\partial y}\frac{\partial}{\partial y}\frac{\partial f}{\partial x} = \frac{\partial^3 f}{\partial y \partial y \partial x} = f_{yyx}$$

or

$$\frac{\partial}{\partial x}\frac{\partial}{\partial y}\frac{\partial}{\partial y}\frac{\partial f}{\partial x} = \frac{\partial^4 f}{\partial x \partial y \partial y \partial x} = \frac{\partial^4 f}{\partial x \partial y^2 \partial x} = f_{xyyx} = f_{xy^2 x}$$

The order of differentiation is kept in the ab-initio fashion in which aliens did commentary on the Calculus of Ibn-ul-Hashem in the latter ages of Islam. Proofs of the Tahir Fundamental Theorem of Partial Differentiation and Tahir Series Expansion for both single and multivariable functions will come later. My proof of the Tahir Series Expansion is a miracle in the history of Calculus that was guarded by the Lord of Muhammed (pbuh) - ALLAH KAREEM – the SUPREME, like a pearl is guarded for a Musalman to come and receive it.

$$\mathbf{f(x) - f(a) = (x - a)f^1(x) - \frac{1}{2!}(x - a)^2 f^2(x)}$$
$$\mathbf{+ \frac{1}{3!}(x - a)^3 f^3(x) - \cdots \infty}$$

For $a = 0$

$$\mathbf{f(x) = f(0) + xf^1(x) - \frac{1}{2!}x^2 f^2(x) + \frac{1}{3!}x^3 f^3(x) - \cdots}$$

$$\mathbf{f(x) = f(0) + \sum_{n=1,2,3}^{n \to \infty} (-1)^{n+1}\frac{1}{n!}x^n f^n(x)}$$

$$\mathbf{f(x, y) = f(0, 0) + (xf_x + yf_y)\big|_{(x,y)}}$$
$$\mathbf{- \frac{1}{2!}(x^2 f_{xx} + 2xyf_{xy} + y^2 f_{yy})\big|_{(x,y)}}$$
$$\mathbf{+ \frac{1}{3!}\Big[x^3 f_{xxx} + x^2 y(f_{yxx} + 2f_{xxy})}$$
$$\mathbf{+ xy^2(f_{xyy} + 2f_{yyx}) + x^3 f_{yyy}\Big]\Big|_{(x,y)} - \cdots}$$

$$\mathbf{f_{yxx} \neq f_{xxy}}$$
$$\mathbf{f_{xyy} \neq f_{yyx}}$$

In both forms of Tahir Series expansion of functions, the signs alternate after the first fixed term of f(0) or f(0,0).

The excellence of Tahir Series expansion of functions over other series is because the partial and ordinary derivatives are all evaluated at (x, y) instead of $(0,0)$. Moreover, Tahir Series expansion of functions is the solution to the fundamental differential equation of Tahironian Cosmos as below:

$$\frac{df(x, y)}{dT} = \frac{\partial f}{\partial x}\frac{dx}{dT} + \frac{\partial f}{\partial y}\frac{dy}{dT}$$

The baby series (macloon & tayloon) don't obey the fundamental differential equation! Tahir series expansion of functions also produces two nice results which eyes have not yet seen before, along with innumerable others, as below:

$$e = \sum_{k=0}^{\infty} \frac{2k+1}{(2k)!}$$

$$e = \sum_{k=0}^{\infty} \frac{2k+2}{(2k+1)!}$$

Putting $y = b$ in the previous equation we see that it reduces to

$$f(x,b) = f(a,b) + \frac{\partial f(x,y)}{\partial x}(x-a) - \frac{1}{2!}\frac{\partial^2 f(x,y)}{\partial x^2}(x-a)^2 + \cdots$$

Now let us define functions $F(x,b)$ & $G(x,b)$ as below:

$$F_1(x,b) = f(x,b) - f(a,b) = \frac{\partial f(x,y)}{\partial x}(x-a) - \frac{1}{2!}\frac{\partial^2 f(x,y)}{\partial x^2}(x-a)^2 + \cdots$$

$$G_1(x,b) = g(x,b) - g(a,b) = \frac{\partial g(x,y)}{\partial x}(x-a) - \frac{1}{2!}\frac{\partial^2 g(x,y)}{\partial x^2}(x-a)^2 + \cdots$$

$F_1(a,b) = 0$ (a peculiar characteristic of the function)

$G_1(a,b) = 0$ (a peculiar characteristic of the function)

We will now form the limit as below:

$$\lim_{x\to a}\frac{F_1(x,b)}{G_1(x,b)} = \frac{0}{0} = \lim_{x\to a}\frac{f(x,b)-f(a,b)}{g(x,b)-g(a,b)} = \lim_{x\to a}\frac{\frac{\partial f(x,y)}{\partial x}(x-a)-\frac{1}{2!}\frac{\partial^2 f(x,y)}{\partial x^2}(x-a)^2 + \cdots}{\frac{\partial g(x,y)}{\partial x}(x-a)-\frac{1}{2!}\frac{\partial^2 g(x,y)}{\partial x^2}(x-a)^2 + \cdots}$$

$$\lim_{x\to a}\frac{F_1(x,b)}{G_1(x,b)} = \frac{0}{0} = \lim_{x\to a}\frac{f(x,b)-f(a,b)}{g(x,b)-g(a,b)} = \lim_{x\to a}\frac{(x-a)\{\frac{\partial f(x,y)}{\partial x}-\frac{1}{2!}\frac{\partial^2 f(x,y)}{\partial x^2}(x-a) + \cdots\}}{(x-a)\{\frac{\partial g(x,y)}{\partial x}-\frac{1}{2!}\frac{\partial^2 g(x,y)}{\partial x^2}(x-a) + \cdots\}}$$

$$\lim_{x\to a}\frac{F_1(x,b)}{G_1(x,b)} = \frac{0}{0} = \lim_{x\to a}\frac{f(x,b)-f(a,b)}{g(x,b)-g(a,b)} = \lim_{x\to a}\frac{\frac{\partial f(x,y)}{\partial x}}{\frac{\partial g(x,y)}{\partial x}}$$

Now from
$$F_1(x,b) = f(x,b) - f(a,b)$$

$$\frac{\partial F_1(x,b)}{\partial x} = \frac{\partial f(x,b)}{\partial x}$$
Similarly

$$\frac{\partial G_1(x,b)}{\partial x} = \frac{\partial g(x,b)}{\partial x}$$

Therefore, we have

$$\lim_{x\to a}\frac{F_1(x,b)}{G_1(x,b)} = \frac{0}{0} = \lim_{x\to a}\frac{\frac{\partial F_1(x,b)}{\partial x}}{\frac{\partial G_1(x,b)}{\partial x}}$$

Now we see that we started from $y = b$, if we now consider the commutative property of Al-jebra by the Grand-Father of Calculus namely Al-Khawarizmi as below:

We see that we have $b = y$, and finally we get:

$$\lim_{x\to a}\frac{F_1(x,y)}{G_1(x,y)} = \frac{0}{0} = \lim_{x\to a}\frac{\frac{\partial F_1(x,y)}{\partial x}}{\frac{\partial G_1(x,y)}{\partial x}}$$

Similarly, by defining the functions $F_2(x,b)$ & $G_2(x,b)$ as below:

$$F_2(x,b) = f(x,b) - f(a,b) - \frac{\partial f(x,y)}{\partial x}(x-a) = -\frac{1}{2!}\frac{\partial^2 f(x,y)}{\partial x^2}(x-a)^2 + \cdots$$

$$G_2(x,b) = g(x,b) - g(a,b) - \frac{\partial g(x,y)}{\partial x}(x-a) = -\frac{1}{2!}\frac{\partial^2 g(x,y)}{\partial x^2}(x-a)^2 + \cdots$$

We can show that:

$$\lim_{x\to a}\frac{F_1(x,y)}{G_1(x,y)} = \frac{0}{0} = \lim_{x\to a}\frac{\frac{\partial^2 F_1(x,y)}{\partial x^2}}{\frac{\partial^2 G_1(x,y)}{\partial x^2}}$$

In general

$$\lim_{\substack{x^1\to a_1\\x^2\to a_2\\x^n\to a_n}}\frac{f(x^1,x^2...x^n)}{g(x^1,x^2...x^n)} = \frac{0}{0} = \lim_{\substack{x^1\to a_1\\x^2\to a_2\\x^n\to a_n}}\frac{\frac{\partial f(x^1,x^2...x^n)}{\partial x^1}}{\frac{\partial g(x^1,x^2...x^n)}{\partial x^1}} = \lim_{\substack{x^1\to a_1\\x^2\to a_2\\x^n\to a_n}}\frac{\frac{\partial^2 f(x^1,x^2...x^n)}{\partial x^1\partial x^1}}{\frac{\partial^2 g(x^1,x^2...x^n)}{\partial x^1\partial x^1}} = \cdots$$

or

$$\lim_{\substack{x^1\to a_1\\x^2\to a_2\\x^n\to a_n}}\frac{f(x^1,x^2...x^n)}{g(x^1,x^2...x^n)} = \frac{0}{0} = \lim_{\substack{x^1\to a_1\\x^2\to a_2\\x^n\to a_n}}\frac{\frac{\partial f(x^1,x^2...x^n)}{\partial x^2}}{\frac{\partial g(x^1,x^2...x^n)}{\partial x^2}} = \lim_{\substack{x^1\to a_1\\x^2\to a_2\\x^n\to a_n}}\frac{\frac{\partial^2 f(x^1,x^2...x^n)}{\partial x^2\partial x^2}}{\frac{\partial^2 g(x^1,x^2...x^n)}{\partial x^2\partial x^2}} = \cdots$$

We will keep on differentiating till we end up finding the limiting value if it exists. But in our scenario, we need to differentiate partially once w.r.t. t_2 as below:

$$\lim_{\substack{x\to 0\\t_1\to T\\t_2\to T}}\frac{x^2 - 2xvt_2^2}{t_1 - t_2} = \frac{0}{0} = \lim_{\substack{x\to 0\\t_1\to T\\t_2\to T}}\frac{\frac{\partial(x^2-2xvt_2^2)}{\partial t_2}}{\frac{\partial(t_1-t_2)}{\partial t_2}} = \lim_{\substack{x\to 0\\t_2\to T}}4xvt_2 = 0$$

Similarly, we have

$$\lim_{\substack{x\to 0\\t_1\to T\\t_2\to T}}\frac{(t_1^2 - t_2^2)}{t_1 - t_2} = \frac{0}{0} = \lim_{\substack{x\to 0\\t_1\to T\\t_2\to T}}\frac{\frac{\partial(t_1^2-t_2^2)}{\partial t_2}}{\frac{\partial(t_1-t_2)}{\partial t_2}} = \lim_{t_2\to T}2t_2 = 2T$$

Now in the limiting case when x tends to zero t_1 & t_2 merge to become T as follows:

Relativistic coordinate time dilation

$$T = \frac{t}{(1 - v^2/c^2)}$$

The time T is independent of both d and the observer going either to the left or to the right above the clock because all the three clocks are coordinate clocks, and their rates of ticking are independent of inertial motion (clocks slowing down in inertial motion). This is the equation for relativistic coordinate time dilation. Time is the happening of an event and if there is no event there is no time. And if there is no definite speed of information (gravitational only and not the speed of light) there is no time dilation. I believe that when ALLAH Kareem – The SUPREME intended to create humans He first created the space for the scene. When HE had created the space and before he created matter in any form and placed it into that space, HE gave space an intrinsic quality and that was that any moving piece of matter or say a clock will undergo slower ticking of time or in other words slowing of time due to the speed of communication s between the two pieces of matter. That quality has been coined the name Gravitivity. That speed of gravitation at the same time was kept equal to another speed of information which is c the speed of light.

$$T = \frac{t}{(1 - v^2/s^2)}$$

In the coming generalized case, as below, of the above experiment we will see how the inertial clock T'comes into being between the above two coordinate clocks namely T and t.

$$T = \frac{T'}{\left(1 - \frac{v^2}{s^2}\right)} = \frac{t}{\left(1 - 2\frac{v^2}{s^2}\right)}$$

In simpler terms we can see that the story of the creation of matter is like the story of the creation of Adam. According to the Scripture of the Heavens and the Earths – Al-Quran, AL-LAH Kareem- The SUPREME said that we created the "dead body" of Adam from mud first. Death was created first, and life was gotten out of it. Then He blew HIS SPIRIT in him, and he became alive. So, the dead body of Adam was created first. When the dead body was created the DNA of life had already been put into it and was waiting for a spirit to function the body at the fullest. The DNA of inertia (gravitivity) without life(gravitation) is the speed of gravitational information s, due to which time slows down in inertial motion! And this is the intrinsic quality of life (gravitation) given to space (dead body).

Gravitivity ≡ DNA(speed of gravitational communication)

Space ≡ Dead body – that contained the DNA.

Gravitation ≡ Spirit

Now let us generalize Tahir Simple Thought Experiment by considering the figure 2 below:

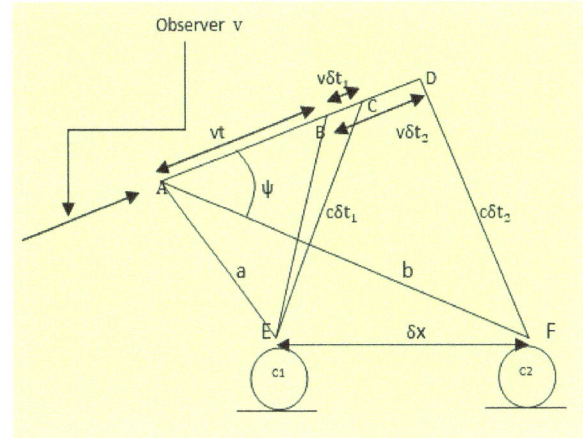

Figure 2. *Relativistic (inertial) time dilation. Deriving the measured time interval T of an inertial clock, but ticking like a coordinate clock in motion, as a limit with the speed of our information in relativistic motion.*

In figure 2 angle EAD is ψ'. Applying Al-Kashi Theorem for Δ EAC we have as below

$$(c\delta t_1)^2 = a^2 + (vt + v\delta t_1)^2 - 2a(vt + v\delta t_1)\cos\psi'$$

Similarly, for Δ FAD we have, by Al-Kashi Theorem, again as below:

$$(c\delta t_2)^2 = b^2 + (vt + v\delta t_2)^2 - 2b(vt + v\delta t_2)\cos\psi$$

Where $T_1 = t + \delta t_1$ & $T_2 = t + \delta t_2$

Subtracting one equation from the other we obtain, after doing some exercise in Al-jebra, as below:

$$c^2(T_2 - t)^2 - c^2(T_1 - t)^2 = b^2 - a^2 + v^2T_2^2 - v^2T_1^2 - 2bvT_2\cos\psi + 2avT_1\cos\psi'$$

Having done some further Algebra, we finally obtain as below:

$$c^2\left(1 - \frac{v^2}{c^2}\right)\frac{(T_2^2 - T_1^2)}{T_2 - T_1} = 2c^2t\frac{(T_2 - T_1)}{T_2 - T_1} + \frac{2avT_1\cos\psi' - 2bvT_2\cos\psi}{T_2 - T_1}$$

In the limiting case when $\delta x \to 0, b \to a, T_2 \to T_1$ & $\psi' \to \psi$ we need to solve three limits as below. We will use Tahir Theorem as developed earlier for finding the limiting values.

$$\lim_{\substack{\delta x \to 0 \\ T_1 \to T \\ T_2 \to T}} \frac{(T_2^2 - T_1^2)}{T_2 - T_1} = \frac{0}{0} = \lim_{\substack{\delta x \to 0 \\ T_1 \to T \\ T_2 \to T}} \frac{\frac{\partial(T_2^2 - T_1^2)}{\partial T_1}}{\frac{\partial(T_2 - T_1)}{\partial T_1}} = \frac{-2T_1}{-1} = 2T$$

$$\lim_{\substack{\delta x \to 0 \\ T_1 \to T \\ T_2 \to T}} \frac{(T_2 - T_1)}{T_2 - T_1} = \frac{0}{0} = \lim_{\substack{\delta x \to 0 \\ T_1 \to T \\ T_2 \to T}} \frac{\frac{\partial(T_2 - T_1)}{\partial T_1}}{\frac{\partial(T_2 - T_1)}{\partial T_1}} = \frac{-1}{-1} = 1$$

&

$$\lim_{\substack{\delta x \to 0 \\ T_1 \to T \\ T_2 \to T \\ \psi \to \psi'}} \frac{(2avT_1\cos\psi' - 2bvT_2\cos\psi)}{T_2 - T_1} = \frac{0}{0} = \lim_{\substack{\delta x \to 0 \\ T_1 \to T \\ T_2 \to T \\ \psi \to \psi'}} \frac{\frac{\partial(2avT_1\cos\psi' - 2bvT_2\cos\psi)}{\partial T_1}}{\frac{\partial(T_2 - T_1)}{\partial T_1}}$$

$$= \lim_{\substack{\delta x \to 0 \\ T_1 \to T \\ T_2 \to T \\ \psi \to \psi'}} \frac{2av\cos\psi'}{-1} = -2av\cos\psi$$

If we redo everything by partially differentiating with respect to T_2 we end up with the same limits, meaning simply that the limits exist. ALLAH does what He pleases (masha-ALLAH). By the Grace of ALLAH KAREEM - the SUPREME, we will find the same results with one clock by a remarkable proof in the miracle article 3!

It is worth mentioning here that the recognition of left or right or towards or away is merely a feeling of recognition of the current position of the moving observer which must be got rid of by the process of taking the limit to make the thought experiment an example of both generality and of the fact that the speed of gravity is a universal constant in space (both inertial and gravitational) and is independent of both the motion of the observer and the source.

So, putting the limiting values we finally obtain

$$T = \frac{t}{\left(1 - \frac{v^2}{c^2}\right)} - \frac{avcos\psi}{c^2\left(1 - \frac{v^2}{c^2}\right)}$$

Replacing a with r we have

$$T = \frac{\left(t - \frac{rvcos\psi}{c^2}\right)}{\left(1 - \frac{v^2}{c^2}\right)} = \frac{T'}{\left(1 - \frac{v^2}{c^2}\right)}$$

$$T\left(1 - \frac{v^2}{c^2}\right) = T'$$

Gravitivity is the scientific study of the speed of gravity. Relativity is the scientific study of the speed of light. Amazing! Isn't it? **T′ is the inertial time of the inertial clock ticking slower in inertial motion. This time T′ is sandwitched between the two times T and t as will be shown in calculations**

coming later in the theory.

$$T = \frac{T'}{\left(1 - \frac{v^2}{s^2}\right)} = \frac{t}{\left(1 - 2\frac{v^2}{s^2}\right)}$$

The speed of gravity is not like the velocity of a ferry in water, a rocket in space or an experimental mirror in motion. It is the speed of an electromagnetic wave.

2.1. World Legalized System (WLS) of Golden Islamic Values (GIVs).

Now let me do some derivations to explain what an imaginary value is and how it is legalized to take the place of real value. The GIVs are the Legalized Values (LVs) which in turn are the imaginary values turned real. How these values turn real has a deep root in the World Religion. In Islam when ALLAH Kareem the Supreme created Adam 1 and Hawwa 1 then he legalized marriage between them through Wahi because in the World Religion Adam is the first Prophet. After marriage the easiest thing in life of the newly married was only one act. When the couple's kids grew up and became adults then the question of the propagation of the sons and daughters of A1 & H1 arose. They already had twins with one boy and a girl. Now to protect them from a forbidden (imaginary) act the LORD legalized marriage between brother and sister only for once between the sons and daughters of A1 & H1 for the propagation of the progeny of the newly created and married couple and hence the race of A1. After that it was cancelled forever! The Lord would also create another couple of A2 & H2 and their kids would have inter-married, and it would not have been difficult for HIM, but he didn't choose that option but rather allowed one girl from one set of twins to be married with the boy from the other set of twins and vice versa. And so, the race propagated. The difference between the two commands of the Lord is like the difference between mathematics and Natural philosophy. In Natural philosophy the same imaginary (forbidden mathematical) values are changed to real values by the decree of the Lord via Tahir. ALLAH Kareem the Supreme does what He pleases and there is no deity to stop him!

To further explain what I have said above, let A(a,b) be the focus of a parabola with directrix line as y = -c. Point B is B(x,-c). Let any arbitrary point on the parabola to be C(x,y). For parabola the eccentricity ratio is 1. So, forming the ratio of two distances, we obtain as below:

$$\frac{\sqrt{(x - a)^2 + (y - b)^2}}{\sqrt{(x - x)^2 + (y + b)^2}} = 1$$

Simplifying, we get

$$y = \frac{x^2 - 2ax + a^2 + b^2 - c^2}{2(c + b)}$$

Now, we will see its intersection with the x axis. So let us put y = 0 in the above equation and solve the quadratic for x. We obtain,

$$x = a \pm \sqrt{c^2 - b^2}$$

Now assume that the parabola doesn't intersect the x axis, i.e., **c < b**

So, we have

$$x = a \pm \pm i\sqrt{b^2 - c^2}$$

The above equation tells Tahir that the quantity $\sqrt{b^2 - c^2}$ was at right angle to the quantity x and it was multiplied with i to bring it back in the direction of x so that it could be added or subtracted from x in the Islamic Algebra. Naïve humanity listened to me very carefully! *The Algebraic quantities are represented on axis x,y & z, so basically, they are directions and cannot be added or subtracted unless or until they are collinear. So, rotations by ±i make them collinear!* So scalar quantities (of the same type) are basically vector quantities because they can be represented by directions of x,y or z!!! There was absolutely no need for commenting on Islamic Physics using vectors. That's why the alien commentary on Islamic Physics has been a colossal failure in the understanding of Islamic Natural Philosophy.

So, the GIVs are as below, by dropping ±i in the above equation.

$$x = a \pm \sqrt{b^2 - c^2}$$

Now let me explain, for the first time in World History, the GIVs as above. They are the imaginary values in mathematics but legalized values in Tahirian Natural Philosophy (TNP) "turned real" but not in mathematics because in mathematics Tahirian Rotations (TR) don't happen. Let us see what they represent on the graph below:

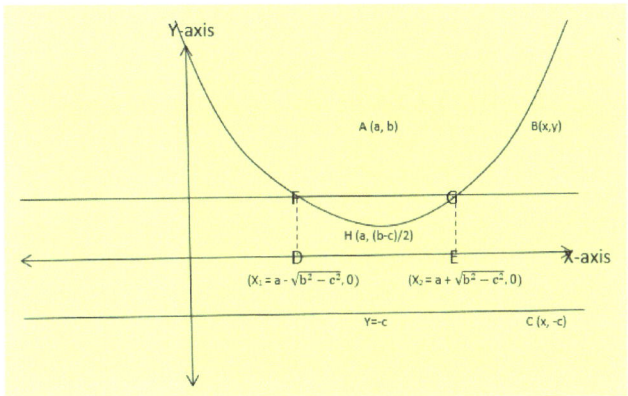

Figure 2.1. *Imaginary values of X_1 & X_2.*

It can be easily shown that the coordinates of the points F & G are ($x_1 = a - \sqrt{b^2 - c^2}$, b-c) & ($x_2 = a + \sqrt{b^2 - c^2}$, $b - c$). Since the x-axis doesn't intersect the parabola, the parabola does what? It measures the shortest distance of the x-axis from itself and finds the points of intersection of a line which intersects the parabola at the same distance away from the same point (turning point of the parabola) at points F

& G. So, the parabola intersects the line at F & G as shown in Figure 2.1 above. Amazing, isn't it? The x coordinates of the points F & G are the imaginary values in mathematics but the legalized values "imaginary turned real" in Tahirian Natural Philosophy (TNP) due to actual rotations of values as will be abundantly shown later.

3. Natural differential equations of inertia and gravitation

Now let us do some derivations for more understanding of Gravitivity and relativity. As Arabic numerals and mathematics are meaningless without the inclusion of zero, gravitation is meaningless without the inclusion of Inertial frame. In other words, Inertia and gravitation are embedded in each other as zero and Arabic Numerals are linked to each other.

For the sake of clarity, we will consider the hypothetical case of inertial body moving with uniform velocity **V** in a straight line as shown in the figure 3 below. We will develop the complete set of Tahirian Differential Equations of Inertial Motion once and forever!

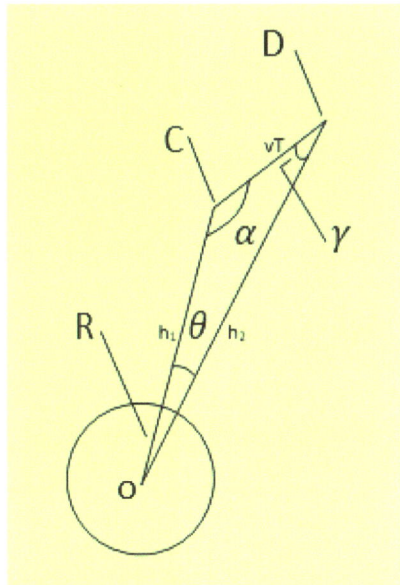

Figure 3. *Tahir universal law of inertia and gravitation − The equivalence of frames.*

We see in the above figure that if we let the body move along the straight path CD for an interval of time T then the distance travelled is vT. Considering triangle OCD and applying Theorem de Al-Kashi, we note that:

$$\overline{OD}^2 = \overline{OC}^2 + \overline{CD}^2 - 2 * \overline{OC} * \overline{CD} * \cos\alpha$$

$$(R + h_2)^2 = (R + h_1)^2 + (VT)^2 - 2(R + h_1)VT\cos\alpha$$

Differentiating the above equation with respect to T′ we obtain

$$(R + h_2)\left(\frac{dh_2}{dT}\right) = 0 + V^2T - (R + h_1)V\cos\alpha$$

$$(R + h_2)\left(\frac{dh_2}{dT}\right) = V^2T - (R + h_1)V\cos\alpha$$

Now putting the limits i.e., when T tends to 0, h_2 tends to h_1 and noting the fact that:

$$r = R + h_1$$

$$\frac{dr}{dT} = \frac{dh_1}{dT}$$

We obtain:

$$r\frac{dr}{dT} = 0 - rV\cos\alpha$$

$$r\left(\frac{dr}{dT}\right) = -rV\cos\alpha$$

$$\frac{dr}{dT} = -V\cos\alpha$$

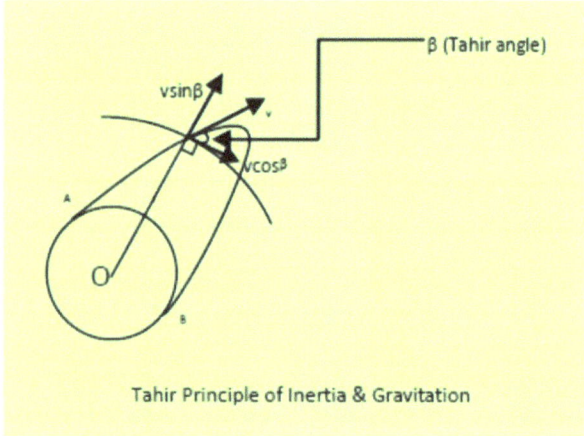

Tahir Principle of Inertia & Gravitation

Figure 4. *Tahir Angle in Tahir Principle of Inertia and Gravitation. Trajectories of Tahirian Projectiles.*

Let me make a change of angle from α to β by the Al-Kashi identity $\alpha = \frac{\pi}{2} + \beta$, thereby getting:

$$\frac{dr}{dT} = V\sin\beta$$

Where $0 \leq \beta \leq 90°$

Actually $\alpha = \frac{\pi}{2} \pm \beta$ for both cases of mass going away from the gravitating mass and mass coming towards the gravitating mass. I have considered the mass going away from the gravitating mass in the above derivation in Figure 3.

Tahir Field Equations

$$\frac{dr}{dT} = v\sin\beta \quad \text{and} \quad r\frac{d\theta}{dT} = v\cos\beta = r\omega$$

Now differentiating the above equation once again the second time with respect to t we obtain

$$(R + h_2)\left(\frac{d^2h_2}{dT^2}\right) + \left(\frac{dh_2}{dT}\right)\left(\frac{dh_2}{dT}\right) = V^2$$

Now putting the limits once again i.e., when T tends to 0, h_2 tends to h_1 and noting the fact again that:

$$r = R + h_1$$

$$\frac{dr}{dT} = \frac{dh_1}{dT}$$

We obtain:

$$r\left(\frac{d^2r}{dT^2}\right) + \left(\frac{dr}{dT}\right)^2 = V^2$$

$$r\left(\frac{d^2r}{dT^2}\right) + (v\sin\beta)^2 = V^2$$

$$\frac{d^2r}{dT^2} = \frac{V^2\cos^2\beta}{r}$$

Tahir Equation *of* Inertia

$$\frac{d^2r}{dT^2} = \frac{V^2\cos^2\beta}{r}$$

Now, in order to derive the third fundamental Tahirian equation of inertial motion we need to consider figure 3 once again as below:

$$\text{Area of triangle OCD} = A = \frac{1}{2}rVT\sin\alpha$$

$$\frac{dA}{dT} = \frac{1}{2}rV\sin\alpha = \frac{h}{2}$$

For uniform motion trajectory of a body h is always a constant in an inertial frame.

$$2\frac{dA}{dT} = rv\sin\alpha = rv\cos\beta = h = constant$$

So, putting together everything of inertial motion in a nutshell we see that the set of three equations of inertia below by Tahir describe our immediate but hypothetical Nature completely. We will see next that the selection of coordinate time T for inertial motion (proper time T′; clocks ticking slow) above was not wrong.

$$\frac{dr}{dT} = v\sin\beta$$

$$\frac{d^2r}{dT^2} - \frac{v^2\cos^2\beta}{r} = 0$$

$$h = rv\cos\beta$$

Where $\alpha = \frac{\pi}{2} + \beta$

We have so far got the two equations as below:

$$\frac{dr}{dT} = v\sin\beta$$

$$\frac{d^2r}{dT^2} - \frac{v^2\cos^2\beta}{r} = 0$$

Where:

$$h = rv\cos\beta$$

We see that if we differentiate the first equation w.r.t. T then we obtain the same equation as below:

$$\frac{d}{dT}\frac{dr}{dT} = \frac{d}{dT}(v\sin\beta) = \frac{d}{dT}\left(\sqrt{v^2 - v^2\cos^2\beta}\right)$$

$$= \frac{d}{dT}\left(\sqrt{v^2 - \frac{h^2}{r^2}}\right)$$

$$\frac{d}{dT}\left(\sqrt{v^2 - \frac{h^2}{r^2}}\right) = \frac{d}{dr}\left(\sqrt{v^2 - \frac{h^2}{r^2}}\right)\frac{dr}{dT} = \frac{v^2\cos^2\beta}{r}$$

The above three Tahir Equations *are* much more sophisticated, elegant and above all meaningful than the entire gravitational physics of my Time! I have basically been able to construct the inertial analogues of gravitation through Tahir's Principle of Instantaneity or *Inter-convertibility*.

Observing the fact that gravitation is the rate of change of velocity, we can substitute acceleration due to gravity in the middle equation *only* as below:

$$\frac{d^2r}{dT^2} - \frac{v^2\cos^2\beta}{r} = -\frac{GM}{r^2}$$

Now we have got a different set of three equations of gravitation as below:

$$\frac{dr}{dT} = v\sin\beta$$

$$\frac{d^2r}{dT^2} - \frac{v^2\cos^2\beta}{r} = -\frac{GM}{r^2}$$

$$h = rv\cos\beta$$

This is *Tahir's Anatomy of Gravitation*. We are now in the Tahirian World of Gravitation in which velocity is no longer a constant inertial velocity but rather a changing gravitational speed! Now, we will insert the first equation in the middle equation and solve it with the third equation differentiated as below:

$$\frac{d}{dT}(v\sin\beta) - \frac{v^2\cos^2\beta}{r} = -\frac{GM}{r^2}$$

$$\frac{d}{dT}(h) = \frac{d}{dT}(rv\cos\beta)$$

$$0 = -rv\sin\beta\frac{d\beta}{dT} + r\cos\beta\frac{dv}{dT} + v\cos\beta\frac{dr}{dT}$$

$$v\cos\beta\frac{d\beta}{dT} + \sin\beta\frac{dv}{dT} - \frac{v^2\cos^2\beta}{r} = -\frac{GM}{r^2}$$

Now solving the two equations for $\frac{dv}{dT}$ & $\frac{d\beta}{dT}$

We obtain as below:

$$\frac{dv}{dT} = -\frac{GM}{r^2}\sin\beta$$

&

$$\frac{d\beta}{dT} = -\frac{GM}{r^2}\frac{\cos\beta}{v} + \frac{v\cos\beta}{r}$$

$$\frac{d\beta}{dT} = [\frac{v}{r} - \frac{1}{v}\frac{GM}{r^2}]\cos\beta$$

Or simply

$$\frac{dv}{dT} = -\frac{GM}{r^2}\sin\beta$$

$$\frac{d\beta}{dT} = [\frac{v}{r} - \frac{1}{v}\frac{GM}{r^2}]\cos\beta$$

Natural Laws remained hid from sight.
ALLAH Kareem wished, Let Tahir Be! And all was light.

Let me prove *remarkably* that the above Tahirian Calculations are perfectly correct by deriving the acceleration i.e., $\frac{dv}{dT}$ of the particle differently as below. We will work in Tahir - Omar-Khayyam's Dynamical-system or Universal System of Dynamics of r, β & θ or simply as **Tahir Universal System of Dynamics (TUSD).** Tahir angle β was of universal importance to be added in the coordinate systems along with the parameter time. Just the time parameter alone was not enough! Hence producing TUSD. Noton etc couldn't introduce Tahir Angle and left their carcasses of physics commentary in miserable condition!

We know for sure that the differential of arc-length exists as below:

$$(ds)^2 = (dr)^2 + (rd\theta)^2$$

Or

$$\left(\frac{ds}{dT}\right)^2 = \left(\frac{dr}{dT}\right)^2 + \left(r\frac{d\theta}{dT}\right)^2$$

We have also come to know through Tahirian Cosmology and TUSD that:

$$v^2 = v^2\sin^2\beta + v^2\cos^2\beta$$

Because

$$\frac{dr}{dT} = v\sin\beta \quad \text{and} \quad r\frac{d\theta}{dT} = v\cos\beta = r\omega$$

Differentiating both sides of the above equation w.r.t. T, I obtain as below.

$$\left(\frac{ds}{dT}\right)^2 = \left(\frac{dr}{dT}\right)^2 + \left(r\frac{d\theta}{dT}\right)^2$$

$$v^2 = \left(\frac{dr}{dT}\right)^2 + \left(r\frac{d\theta}{dT}\right)^2$$

$$v\frac{dv}{dT} = \frac{dr}{dT}\frac{d(\frac{dr}{dT})}{dT} + r\frac{d\theta}{dT}\frac{d(r\frac{d\theta}{dT})}{dT}$$

$$v\frac{dv}{dT} = \frac{dr}{dT}\frac{d^2r}{dT^2} + r\frac{d\theta}{dT}\left[r\frac{d^2\theta}{dT^2} + \frac{dr}{dT}\frac{d\theta}{dT}\right]$$

$$v\frac{dv}{dT} = \frac{dr}{dT}\frac{d^2r}{dT^2} + r^2\frac{d\theta}{dT}\frac{d^2\theta}{dT^2} + r\frac{dr}{dT}\left[\frac{d\theta}{dT}\right]^2$$

Now from here onward, I will play a trick. I will add and

subtract the term $2r\frac{dr}{dT}\left[\frac{d\theta}{dT}\right]^2$ as below:

$$v\frac{dv}{dT} = \frac{dr}{dT}\frac{d^2r}{dT^2} + r^2\frac{d\theta}{dT}\frac{d^2\theta}{dT^2} + r\frac{dr}{dT}\left[\frac{d\theta}{dT}\right]^2 - 2r\frac{dr}{dT}\left[\frac{d\theta}{dT}\right]^2 + 2r\frac{dr}{dT}\left[\frac{d\theta}{dT}\right]^2$$

Rearranging terms, Tahir obtains as below:

$$v\frac{dv}{dT} = \frac{dr}{dT}\frac{d^2r}{dT^2} + r\frac{dr}{dT}\left[\frac{d\theta}{dT}\right]^2 - 2r\frac{dr}{dT}\left[\frac{d\theta}{dT}\right]^2 + r^2\frac{d\theta}{dT}\frac{d^2\theta}{dT^2} + 2r\frac{dr}{dT}\left[\frac{d\theta}{dT}\right]^2$$

Finally, in the History of Islamic Physics, I obtain as below:

$$\frac{dv}{dT}v = \left[\frac{d^2r}{dT^2} - r\left[\frac{d\theta}{dT}\right]^2\right]\frac{dr}{dT} + \left[r\frac{d^2\theta}{dT^2} + \frac{dr}{dT}\frac{d\theta}{dT} + \frac{dr}{dT}\frac{d\theta}{dT}\right]r\frac{d\theta}{dT}$$

This Equation of TUSD is simply the Equation of Every Dynamic/Mechanical or Electro-Mechanical System like gravity!

We can also write the equation as

$$\frac{dv}{dT} = \left[\frac{d^2r}{dT^2} - r\left[\frac{d\theta}{dT}\right]^2\right]\sin\beta + \left[r\frac{d^2\theta}{dT^2} + \frac{dr}{dT}\frac{d\theta}{dT} + \frac{dr}{dT}\frac{d\theta}{dT}\right]\cos\beta$$

This is the Equation of all motion.

Tahir Field Equation *of* all Motion

$$\frac{dv}{dT} = \left[\frac{d^2r}{dT^2} - r\left[\frac{d\theta}{dT}\right]^2\right]\sin\beta + \left[r\frac{d^2\theta}{dT^2} + \frac{dr}{dT}\frac{d\theta}{dT} + \frac{dr}{dT}\frac{d\theta}{dT}\right]\cos\beta$$

Very important thing to remember here is that $\sin\beta$ and $\cos\beta$ are the terms of the polar metric and *NOT* the terms taken from inertial motion (non-accelerated frame of reference) conditions or non-inertial motion (accelerated frame of reference) conditions! General Field Equations or Tahirian Field Equations are, as evident from their names, general equations of *all* motion and have nothing to do with non-accelerated or accelerated motion. *Amazing, isn't it?* We derive the non-accelerated and accelerated motion equations after putting their respective conditions in the Tahirian Field Equations! Basically, it's the Tahirian Universal System of Dynamics (TUSD) that describes Tahironian (Universal) Motion.

Let me find the partner equation as below:

$$\frac{dr}{dT} = V\sin\beta \quad \text{and} \quad r\frac{d\theta}{dT} = v\cos\beta$$

$$\tan\beta = \frac{\frac{dr}{dT}}{r\frac{d\theta}{dT}}$$

Differentiating both sides of the above equation w.r.t. T, I obtain as below.

$$\frac{d(\tan\beta)}{dT} = \frac{d}{dT}\left[\frac{\frac{dr}{dT}}{r\frac{d\theta}{dT}}\right]$$

$$\sec^2\beta \frac{d\beta}{dT} = \frac{v\cos\beta \frac{d}{dT}\left[\frac{dr}{dT}\right] - \frac{dr}{dT}\frac{d}{dT}\left[r\frac{d\theta}{dT}\right]}{v^2\cos^2\beta}$$

$$\frac{1}{\cos^2\beta}\frac{d\beta}{dT} = \frac{v\cos\beta \frac{d}{dT}\left[\frac{dr}{dT}\right] - \frac{dr}{dT}\left[r\frac{d^2\theta}{dT^2} + \frac{dr}{dT}\frac{d\theta}{dT}\right]}{v^2\cos^2\beta}$$

$$\frac{d\beta}{dT} = \frac{v\cos\beta \frac{d}{dT}\left[\frac{dr}{dT}\right] - \frac{dr}{dT}\left[r\frac{d^2\theta}{dT^2} + \frac{dr}{dT}\frac{d\theta}{dT}\right]}{v^2}$$

$$\frac{d\beta}{dT} = \frac{v\cos\beta \frac{d^2r}{dT^2} - v\sin\beta\left[r\frac{d^2\theta}{dT^2} + \frac{dr}{dT}\frac{d\theta}{dT}\right]}{v^2}$$

$$\frac{d\beta}{dT} = \frac{d^2r}{dT^2}\frac{\cos\beta}{v} - \left[r\frac{d^2\theta}{dT^2} + \frac{dr}{dT}\frac{d\theta}{dT}\right]\frac{\sin\beta}{v}$$

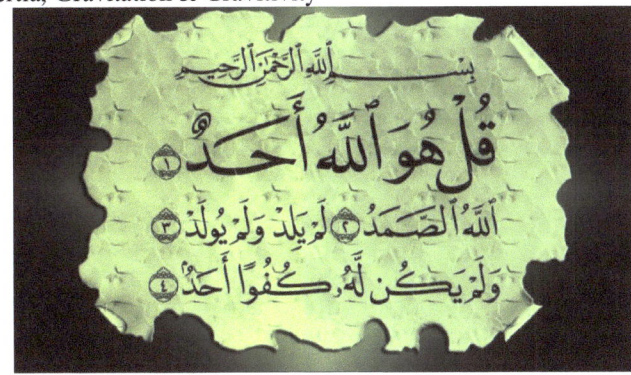

In the name of ALLAH, the most Gracious, the most Merciful.

Say: He is ALLAH Kareem – The Supreme, the One and Only. ALLAH Kareem – the Supreme, the Eternal, the Absolute. He begets not, nor is He begotten. And there is none like unto Him.

Equation *of* all Motion

$$\frac{d\beta}{dT} = \frac{1}{V}\frac{d^2r}{dT^2}\cos\beta - \frac{1}{V}\left[r\frac{d^2\theta}{dT^2} + \frac{dr}{dT}\frac{d\theta}{dT}\right]\sin\beta$$

So, the general equations of motion or simply Tahir Field Equations are as below:

Tahir Field Equations or General Equations of all motion

$$\frac{dV}{dT} = \left[\frac{d^2r}{dT^2} - r\left[\frac{d\theta}{dT}\right]^2\right]\sin\beta + \left[r\frac{d^2\theta}{dT^2} + \frac{dr}{dT}\frac{d\theta}{dT} + \frac{dr}{dT}\frac{d\theta}{dT}\right]\cos\beta$$

$$\frac{d\beta}{dT} = \frac{1}{V}\frac{d^2r}{dT^2}\cos\beta - \frac{1}{V}\left[r\frac{d^2\theta}{dT^2} + \frac{dr}{dT}\frac{d\theta}{dT}\right]\sin\beta$$

A remarkable beauty of these equations is that they come even before inertia steps in! No one believed and knew that there could be a set of equations even before inertia except Tahir. Everyone thought that inertia is foremost and the only to precede gravitation, but this is not the case. ALLAH-o-Akbar! They constitute the deepest known connection between Tahirian physics and Tahirian mathematics of all times.

Now let me derive the equations of inertia from the above parent equations as below.

$$\frac{dv}{dT} = 0 \text{ (Very easy)}$$

$$\frac{d\beta}{dT} = \frac{v\cos\beta}{r}$$

The first equation is straightforward. We can set the two terms equal to zero as below to obtain the first desired equation.

$$\frac{d^2r}{dT^2} - r\left[\frac{d\theta}{dT}\right]^2 = 0 \ \& \ \overline{r\frac{d^2\theta}{dT^2} + \frac{dr}{dT}\frac{d\theta}{dT} + \frac{dr}{dT}\frac{d\theta}{dT}} = 0$$

Now substituting the above conditions of Tahirian Inertia back into the second Tahir Field equation, we obtain the second equation of Tahirian Inertia as above. Similarly with a little Tahirian Intellect, we can obtain the gravitational derivatives as below:

$$\frac{dV}{dT} = -\frac{GM}{r^2}\sin\beta$$

$$\frac{d\beta}{dT} = \left[\frac{v}{r} - \frac{GM}{vr^2}\right]\cos\beta$$

Tahirian Gravitation is the derivative of Tahirian Inertia and Tahirian Inertia is the derivative of Tahir Field Equations!

Now let Tahir do further remarkable "Anatomy *of* Inertia" in the Islamic History once and forever by using the above equation as below:

$$\frac{dv}{dT} = \left[\frac{d^2r}{dT^2} - r\left[\frac{d\theta}{dT}\right]^2\right]\sin\beta + \left[r\frac{d^2\theta}{dT^2} + \frac{dr}{dT}\frac{d\theta}{dT} + \frac{dr}{dT}\frac{d\theta}{dT}\right]\cos\beta$$

For inertial motion in a straight line $\frac{dv}{dT} = 0$ as per the first law of motion by the Father of Inertia, Ibn e Sina.

$$0 = [\frac{d^2r}{dT^2} - r[\frac{d\theta}{dT}]^2]\sin\beta + \overline{[r\frac{d^2\theta}{dT^2} + \frac{dr}{dT}\frac{d\theta}{dT} + \frac{dr}{dT}\frac{d\theta}{dT}]}\cos\beta$$

Tahir is the second Ibne-Sina.

This gives us the following two differential equations of Tahir Scale Importance (TSI) in Islamic Physics of the Principles of Inertia as below:

$$\frac{d^2r}{dT^2} - r[\frac{d\theta}{dT}]^2 = 0 \qquad \textbf{Tahir Equation One (TEO)}$$

$$\&$$

$$r\frac{d^2\theta}{dT^2} + 2\frac{dr}{dT}\frac{d\theta}{dT} = 0 \qquad \textbf{Tahir Equation Two (TET)}$$

They can also be called the Equations of Tahirian Cosmology.

Solving TET, we obtain as below:

$$r\frac{d^2\theta}{dT^2} + 2\frac{dr}{dT}\frac{d\theta}{dT} = 0$$

$$r\frac{d}{dT}[\frac{d\theta}{dT}] + 2\frac{dr}{dT}\frac{d\theta}{dT} = 0$$

$$r\frac{d}{dT}[\omega] + 2\frac{dr}{dT}\frac{d\theta}{dT} = 0$$

$$r\frac{d}{dr}[\omega]\frac{dr}{dT} + 2\frac{dr}{dT}\frac{d\theta}{dT} = 0$$

$$[r\frac{d}{dr}[\omega] + 2\frac{d\theta}{dT}]\frac{dr}{dT} = 0$$

$$r\frac{d}{dr}[\omega] + 2\frac{d\theta}{dT} = 0$$

$$r\frac{d\omega}{dr} + 2\omega = 0$$

$$\frac{d\omega}{\omega} = -2\frac{dr}{r}$$

$$\int\frac{d\omega}{\omega} = -2\int\frac{dr}{r}$$

$$\ln\omega = -2\ln r + C'$$

$$\text{Let } C' = \ln C$$

$$\ln\omega = \ln r^{-2} + \ln C$$

$$\ln\omega = \ln Cr^{-2}$$

$$\omega = Cr^{-2}$$

$$\omega = \frac{C}{r^2}$$

$$\frac{v\cos\beta}{r} = \frac{C}{r^2}$$

$$C = rv\cos\beta = \omega r^2$$

Let this constant of integration C be called h. Do you know of anyone who has found a more meaningful constant of integration than the Tahir Constant of Integration in World Physics?

$$h = rv\cos\beta = \omega r^2$$

Writing TEO again as below

$$\frac{d^2r}{dT^2} - r[\frac{d\theta}{dT}]^2 = 0$$

$$\frac{d^2r}{dT^2} - \frac{v^2\cos^2\beta}{r} = 0$$

$$\&$$

$$h = rv\cos\beta$$

We obtain the Tahirian Solution (TS) of the above differential equation, for the first time in history, as

$$\frac{dr}{dT} = \pm\pm iV\cos\beta$$

The above value states that the component $V\cos\beta$ of velocity V is rotated 90° in the direction of $\frac{dr}{dT}$. Since this rotation is meant to be as real as anything in the universe, so it will be considered as a part of Legalized System of Values of Golden Islamic Values. So, we will drop $\pm i$ (rotation clockwise and anti-clockwise) from the above equation and consider the value as below:

$$\frac{dr}{dT} = \pm V\cos\beta$$

where $i = \pm\sqrt{-1}$ (rotation of 90°) & $h = rv\cos\beta$

$$\frac{d^2r}{dT^2} = \frac{V^2\cos^2\beta}{r} = r[\frac{d\theta}{dT}]^2 = \omega^2 r$$

Now let me do Tahir Anatomy of Inertia for the first time in World History as below using TET:

$$r\frac{d^2\theta}{dT^2} + 2\frac{dr}{dT}\frac{d\theta}{dT} = 0$$

Or in Tahirian Reality, as below

$$\overline{r\frac{d^2\theta}{dT^2} + \frac{dr}{dT}\frac{d\theta}{dT} + \frac{dr}{dT}\frac{d\theta}{dT}} = 0$$

Using the Colossal Equations of Tahirian Cosmology, we have as below:

$$\frac{dr}{dT} = \pm iV\cos\beta \text{ and } r\frac{d\theta}{dT} = v\cos\beta \text{ \& } \frac{dr}{dT} = \pm V\sin\beta$$

Or just

$$\frac{dr}{dT} = iV\cos\beta \text{ and } r\frac{d\theta}{dT} = v\cos\beta \text{ \& } \frac{dr}{dT} = V\sin\beta$$

In order, to find the important values, we need to differentiate the second equation once again as below:

$$v\cos\beta = r\frac{d\theta}{dT}$$

$$\frac{d(v\cos\beta)}{dT} = \frac{d(r\frac{d\theta}{dT})}{dT}$$

$$\frac{d(v\cos\beta)}{dT} = r\frac{d^2\theta}{dT^2} + \frac{dr}{dT}\frac{d\theta}{dT}$$

$$\frac{d(v\cos\beta)}{dT} = r\frac{d^2\theta}{dT^2} + \frac{dr}{dT}\frac{d\theta}{dT}$$

$$\frac{d(v\cos\beta)}{dT} = r\frac{d}{dT}\frac{d\theta}{dT} + \frac{dr}{dT}\frac{d\theta}{dT}$$

$$\frac{d(v\cos\beta)}{dT} = r\frac{d}{dT}\left(\frac{v\cos\beta}{r}\right) + \frac{dr}{dT}\frac{d\theta}{dT}$$

$$\frac{d(v\cos\beta)}{dT} = r\frac{d}{dT}\left(\frac{h}{r^2}\right) + \frac{dr}{dT}\frac{d\theta}{dT}$$

$$\frac{d(v\cos\beta)}{dT} = r\times -2\frac{h}{r^3}\frac{dr}{dT} + \frac{dr}{dT}\frac{d\theta}{dT}$$

$$\frac{d(v\cos\beta)}{dT} = -2\frac{v\cos\beta}{r}\frac{dr}{dT} + \frac{d\theta}{dT}\frac{dr}{dT}$$

$$\frac{d(v\cos\beta)}{dT} = \left[-2\frac{v\cos\beta}{r} + \frac{v\cos\beta}{r}\right]\frac{dr}{dT}$$

$$\frac{d(v\cos\beta)}{dT} = -\frac{v\cos\beta}{r}\frac{dr}{dT} = -\frac{d\theta}{dT}\frac{dr}{dT}$$

$$\frac{d(v\cos\beta)}{dT} = -\frac{v\cos\beta}{r}\times \pm iV\cos\beta$$

$$\frac{d(v\cos\beta)}{dT} = -\pm i\frac{V^2\cos^2\beta}{r}$$

$$\overline{r\frac{d^2\theta}{dT^2} + \frac{dr}{dT}\frac{d\theta}{dT} - \frac{d(v\cos\beta)}{dT}} = 0$$

I have already shown before that if I differentiate w.r.t. T then I obtain:

$$\frac{d^2r}{dT^2} = \frac{d}{dT}\frac{dr}{dT} = \frac{d}{dT}(v\sin\beta)$$

$$= \frac{d}{dT}\left(\sqrt{v^2 - v^2\cos^2\beta}\right)$$

$$= \frac{d}{dT}\left(\sqrt{v^2 - \frac{h^2}{r^2}}\right)$$

$$\frac{d}{dT}\left(\sqrt{v^2 - \frac{h^2}{r^2}}\right) = \frac{d}{dr}\left(\sqrt{v^2 - \frac{h^2}{r^2}}\right)\frac{dr}{dT} = \frac{v^2\cos^2\beta}{r}$$

$$\frac{d}{dT}(v\sin\beta) = \frac{v^2\cos^2\beta}{r} = \frac{d^2r}{dT^2}$$

$$-\frac{d(v\cos\beta)}{dT} = \pm i\frac{V^2\cos^2\beta}{r} = \frac{dr}{dT}\frac{d\theta}{dT}$$

So, we see the similarity as below:

$$\overline{r\frac{d^2\theta}{dT^2} + \frac{dr}{dT}\frac{d\theta}{dT} - \frac{d(v\cos\beta)}{dT}} = 0$$

$$\frac{d^2r}{dT^2} - \frac{d(v\sin\beta)}{dT} = 0$$

The above will be substituted in my equation gone previously for Inertia as below:

$$0 = \left[\frac{d^2r}{dT^2} - \frac{d(v\sin\beta)}{dT}\right]\sin\beta + \left[\overline{r\frac{d^2\theta}{dT^2} + \frac{dr}{dT}\frac{d\theta}{dT} - \frac{d(v\cos\beta)}{dT}}\right]\cos\beta$$

$$0 = \left[\frac{d(v\sin\beta)}{dT} - \frac{d(v\sin\beta)}{dT}\right]\sin\beta + \left[\frac{d(v\cos\beta)}{dT} - \frac{d(v\cos\beta)}{dT}\right]\cos\beta$$

Now it is the time for the destiny of the sons and daughters of Adam and Hawwa to change forever as below:

$$\frac{dv}{dT} = \left[\frac{d^2r}{dT^2} - r\left[\frac{d\theta}{dT}\right]^2\right]\sin\beta + \left[r\frac{d^2\theta}{dT^2} + \frac{dr}{dT}\frac{d\theta}{dT}\right]\cos\beta$$

Writing the above equation, a bit differently as below

$$\frac{dv}{dT} = A\sin\beta + B\cos\beta$$

In the above equation A & B are both equal to zero for inertia but for pure gravitation only B is equal to zero. But very interestingly when B tends to zero for gravitation the $\frac{dv}{dT}$ vector doesn't fall on A. ALLAH-O-AKBAR!!! It still serves as the *resultant in the same direction* but with a smaller magnitude this time. Why because the B doesn't vanish in Tahirian Reality but rather splits into inertial couples which are still present in pure gravitation. Amazing, isn't it?

Hence the Tahirian Resultant

$$\frac{dv}{dT} = A\sin\beta$$

$$B\cos\beta = Payload$$

However, the payload is to be released by both inertia and gravitation. We will see it later.

Usually when one rectangular component becomes equal to zero then the resultant vector falls on to the other rectangular component as below:

$$\frac{dv}{dT} = \sqrt{A^2 + B^2} \rightarrow \frac{dv}{dT} = A \text{ when } B \rightarrow 0$$

So, the Tahirian Resultant is a Projected Resultant in which the direction of the resultant doesn't change even when one force is *partially* removed because it exits in the form of Tahirian Inertial Couple(s). Whereas in ordinary resultant vector when one component force is *wholly* removed, and it doesn't exist in any form anymore, then the resultant changes its direction and magnitude both. What I mean to say is that in Tahirian Cosmology gravitational force is never ever the resultant force and the Tahirian Projected Resultant $\frac{dv}{dT}$ is the net force or resultant and is the active component of gravity in the direction of motion after the payload is released.

As an example of the above, consider a right-angle triangle ABC with vertices A, B & C and the corresponding sides as a, b & c. Angle C is right-angle. We know from Islamic Trigonometry that the side C can be expressed as below in two different ways.

$$c = \sqrt{a^2 + b^2}$$

$$c = a\sin B + b\sin A$$

Now, if we substitute b = 0 in the above equations, we see that magnitude of c changes in both, but direction changes in only one of them. One is the ordinary resultant, and the other is the projected resultant or Tahirian Resultant.

If I substitute

$$\frac{dr}{dT} = \pm iV\cos\beta$$

Then I have

$$\left[\frac{d(\pm iv\cos\beta)}{dT} - \frac{d(\pm iv\cos\beta)}{dT}\right]\sin\beta + \left[\pm i\frac{V^2\cos^2\beta}{r} - \pm i\frac{V^2\cos^2\beta}{r}\right]\cos\beta = 0$$

$$\left[\frac{d(\pm iv\cos\beta)}{dT} - \frac{d(\pm iv\cos\beta)}{dT}\right]\sin\beta + \left[\frac{d(\pm iv\cos\beta)}{dT} - \frac{d(\pm iv\cos\beta)}{dT}\right]\cos\beta = 0$$

$$\left[\frac{d(v\sin\beta)}{dT} - \frac{d(v\sin\beta)}{dT}\right]\sin\beta + \left[\frac{d(\pm iv\cos\beta)}{dT} - \frac{d(\pm iv\cos\beta)}{dT}\right]\cos\beta = 0$$

The above equation is one equation and not two even though we have double signs, see as below:

$$\left[\frac{d^2r}{dT^2} - \frac{V^2\cos^2\beta}{r}\right]\sin\beta + \left[-\pm i\frac{V^2\cos^2\beta}{r} + \pm i\frac{V^2\cos^2\beta}{r}\right]\cos\beta = 0$$

$$i^2\left[\frac{V^2\cos^2\beta}{r} - \frac{d^2r}{dT^2}\right]\sin\beta + \left[+ \pm i\frac{V^2\cos^2\beta}{r} - \pm i\frac{V^2\cos^2\beta}{r}\right]\cos\beta = 0$$

The terms in brackets in the above two equations are the same! *Two equations must be different in at least one term to be added together in Tahirian Reality.*

Humanity! now lend me your ears. Looking, at the above equation, with Tahir Standard Consciousness (TSC), we see that there is a couple of Inertial Acceleration as below:

$$\pm i\frac{V^2\cos^2\beta}{r}$$

I will drop $\pm i$ terms because they are just to show that the

quantity, that has been previously *normal* to the plane swept out by the line joining the Centre of rotation and the particle in Mechanical motion under a mechanical system, has been rotated by 90^0 into the same plane and collinear with the $\frac{dv}{dT}$ term or a quantity has undergone a rotation of 180 degrees as below

$$i^2\left[\frac{V^2\cos^2\beta}{r} - \frac{d^2r}{dT^2}\right]$$

ALLAH Kareem the Supreme has created gravity as a special mechanical system for us and for the particles Jupiter, Saturn, Sun etc. as well.

So, we can write as

$$\frac{V^2\cos^2\beta}{r} - \frac{d^2r}{dT^2} \quad \& \quad \frac{V^2\cos^2\beta}{r}$$

They are called the Tahirian Inertial Accelerations,

They are also called the Fundamental Accelerations of TUSD.

The Tahirian Centripetal/Centrifugal Forces (TCCFs) couple is as below:

$$i^2 m\left[\frac{V^2\cos^2\beta}{r} - \frac{d^2r}{dT^2}\right]$$

lie in the horizontal plane XY containing the Tahirian parameters of **r, θ & β**. Whereas the couple of Tahirian Inertial Acceleration as below:

$$\pm\frac{V^2\cos^2\beta}{r}$$

lie normal to the horizontal plane XY containing the parameters **r, θ & β**. *Whenever the TCCFs are replaced by a mechanical or physical acceleration like gravity or gravity induced forces like tension, compression or reactions at the supports etc. The Tahirian Centripetal Force is 180° apart from the Tahirian Centrifugal Force! Tahirian Centripetal and centrifugal forces are both inertial and come into action as a result to oppose the centripetal and centrifugal mechanical or physical forces like compression or tension in structural elements under dynamic action. I will discuss scenarios in which inertial centrifugal and centripetal forces come into action to oppose the mechanical centripetal and centrifugal forces like compression or tension in strings or members in dynamics.*

Another couple is as below:

$$\pm\frac{V^2\cos\beta\sin\beta}{r}$$

Similarly, we can repeat the above calculations with $\frac{dr}{dT} =$

$V\sin\beta$ and obtain:

$$\overline{r\frac{d^2\theta}{dT^2} + \frac{dr}{dT}\frac{d\theta}{dT}} = -\frac{V^2\cos\beta\sin\beta}{r}$$

&

$$\left[\frac{d^2r}{dT^2} - r\left[\frac{d\theta}{dT}\right]^2\right]\sin\beta + \left[r\frac{d^2\theta}{dT^2} + 2\frac{dr}{dT}\frac{d\theta}{dT}\right]\cos\beta = 0$$

Therefore, the above equation can be written as below:

$$\frac{dv}{dT} = \left[\frac{d^2r}{dT^2} - r\left[\frac{d\theta}{dT}\right]^2\right]\sin\beta + \left[+\frac{V^2\cos\beta\sin\beta}{r} - \frac{V^2\cos\beta\sin\beta}{r}\right]\cos\beta = 0$$

So, we can eventually write as below:

$$\frac{V^2\cos^2\beta}{r} - \frac{d^2r}{dT^2}, \quad \frac{V^2\cos^2\beta}{r} \quad \& \quad \frac{V^2\cos\beta\sin\beta}{r}$$

They are called the Tahirian Inertial Accelerations,

They are also called the Fundamental Accelerations of TUSD.

When ALLAH Kareem – the Supreme – created the space, and before he placed matter into it, he gave the space the above inertial accelerations.

These are the Tahirian Signatures of Space & Time!

I am establishing the equations of Tahirian Inertia or Tahirian Inertial Motion or non-Accelerated Motion out of the Tahir Field Equations or the Tahirian General Equations of motion by applying the conditions of Tahirian Inertia i.e., the conditions of non-accelerated frame of reference. Since the terms in brackets are totally different in the above equations, they can be added together as below:

$$\frac{dv}{dT} = \left[\frac{d^2r}{dT^2} - r\left[\frac{d\theta}{dT}\right]^2\right]\sin\beta + \left[-\frac{V^2\cos\beta\sin\beta}{r} + \frac{V^2\cos\beta\sin\beta}{r} - i\frac{V^2\cos^2\beta}{r} + i\frac{V^2\cos^2\beta}{r}\right]\cos\beta = 0$$

The above equation can also be written as below after dropping i

$$\frac{dv}{dT} = \left[\frac{d^2r}{dT^2} - r\left[\frac{d\theta}{dT}\right]^2\right]\sin\beta + \left[-\frac{V^2\cos\beta\sin\beta}{r} + \frac{V^2\cos\beta\sin\beta}{r} - \frac{V^2\cos^2\beta}{r} + \frac{V^2\cos^2\beta}{r}\right]\cos\beta = 0$$

In Tahirian Natural Philosophy, we don't take average or mean unless or until we talk about measured data. Tahirian Natural Philosophy is the philosophy of Islamic Science of the highest intellect, i.e.,

If c = a & c is also equal to b then c = a + b

Now let us find the partner equation for inertia as below.

Utilizing, the Tahir Field Equation, we have

$$\frac{d\beta}{dT} = \frac{d^2r}{dT^2}\frac{\cos\beta}{v} - \left[r\frac{d^2\theta}{dT^2} + \frac{dr}{dT}\frac{d\theta}{dT}\right]\frac{\sin\beta}{v}$$

By putting the conditions of inertia, we obtain the equations as below:

$$\frac{d\beta}{dT} = \frac{d^2r}{dT^2}\frac{\cos\beta}{v} - \left[-2\frac{V^2\cos\beta\sin\beta}{r} + \frac{V^2\cos\beta\sin\beta}{r}\right]\frac{\sin\beta}{v}$$

$$\frac{d\beta}{dT} = \frac{d^2r}{dT^2}\frac{\cos\beta}{v} + \frac{V^2\cos\beta\sin\beta}{r}\frac{\sin\beta}{v}$$

&

$$\frac{d\beta}{dT} = \left[\frac{d^2r}{dT^2}\right]\frac{\cos\beta}{v} - \left[-\pm 2i\frac{V^2\cos^2\beta}{r} + \pm i\frac{V^2\cos^2\beta}{r}\right]\frac{\sin\beta}{v}$$

Brackets have been added above for clarity.

$$\frac{d\beta}{dT} = \left[\frac{d^2r}{dT^2}\right]\frac{\cos\beta}{v} - \left[-\pm i\frac{V^2\cos^2\beta}{r}\right]\frac{\sin\beta}{v}$$

$$\frac{d\beta}{dT} = \left[\frac{d^2r}{dT^2}\right]\frac{\cos\beta}{v} + \left[\pm i\frac{V^2\cos^2\beta}{r}\right]\frac{\sin\beta}{v}$$

Now, we have two *different* equations as below:

$$\frac{d\beta}{dT} = \left[\frac{d^2r}{dT^2}\right]\frac{\cos\beta}{v} + \frac{V^2\cos\beta\sin\beta}{r}\frac{\sin\beta}{v}$$

$$\frac{d\beta}{dT} = \left[\frac{d^2r}{dT^2}\right]\frac{\cos\beta}{v} + \left[\pm i\frac{V^2\cos^2\beta}{r}\right]\frac{\sin\beta}{v}$$

The above two equations must produce one equation as below:

$$\frac{d\beta}{dT} = \left[\frac{d^2r}{dT^2}\right]\frac{\cos\beta}{v} + \left[\frac{V^2\cos\beta\sin\beta}{r} \pm i\frac{V^2\cos^2\beta}{r}\right]\frac{\sin\beta}{v}$$

Now the above equation can again be split up into two *different* equations as below:

$$\frac{d\beta}{dT} = \left[\frac{d^2r}{dT^2}\right]\frac{\cos\beta}{v} + \left[\frac{V^2\cos\beta\sin\beta}{r} + i\frac{V^2\cos^2\beta}{r}\right]\frac{\sin\beta}{v}$$

$$\frac{d\beta}{dT} = \left[\frac{d^2r}{dT^2}\right]\frac{\cos\beta}{v} + \left[\frac{V^2\cos\beta\sin\beta}{r} - i\frac{V^2\cos^2\beta}{r}\right]\frac{\sin\beta}{v}$$

The above two equations must produce one equation as below after Tahirian Addition of Natural Philosophy

$$\frac{d\beta}{dT} = \left[\frac{d^2r}{dT^2}\right]\frac{\cos\beta}{v} + \left[\frac{V^2\cos\beta\sin\beta}{r}\right]\frac{\sin\beta}{v}$$

Or

$$\frac{d\beta}{dT} = \left[\frac{V^2\cos^2\beta}{r}\right]\frac{\cos\beta}{v} + \left[\frac{V^2\cos\beta\sin\beta}{r}\right]\frac{\sin\beta}{v}$$

$$\frac{d\beta}{dT} = \frac{v\cos\beta}{r}(\cos^2\beta + \sin^2\beta)$$

Or

$$\frac{d\beta}{dT} = \frac{v\cos\beta}{r}$$

The game of $\frac{dr}{dT} = \pm iV\cos\beta$ is for terms within the square brackets only!

So, the equations of Tahirian Inertial Motion (TIM) are:

$$\frac{dv}{dT} = [\frac{d^2r}{dT^2} - r[\frac{d\theta}{dT}]^2]\sin\beta + [-\frac{V^2\cos\beta\sin\beta}{r} + \frac{V^2\cos\beta\sin\beta}{r} - i\frac{V^2\cos^2\beta}{r} + i\frac{V^2\cos^2\beta}{r}]\cos\beta = 0$$

Or, after dropping i as per the laws of the World legalized System of GIVs

$$\frac{dv}{dT} = [\frac{d^2r}{dT^2} - r[\frac{d\theta}{dT}]^2]\sin\beta + [-\frac{V^2\cos\beta\sin\beta}{r} + \frac{V^2\cos\beta\sin\beta}{r} - \frac{V^2\cos^2\beta}{r} + \frac{V^2\cos^2\beta}{r}]\cos\beta = 0$$

$$\frac{d\beta}{dT} = \frac{v\cos\beta}{r}$$

$$\frac{V^2\cos^2\beta}{r} - \frac{d^2r}{dT^2}, \quad \frac{V^2\cos^2\beta}{r} \quad \& \quad \frac{V^2\cos\beta\sin\beta}{r}$$

But listen, the above is just a Tahirian Triad *of* Inertial Accelerations that I will extensively use in solving the unsolved problems in Islamic History for the last 1000 years.

After removing the payload from the above equation, it reduces to

$$\frac{dv}{dT} = \left\{ \frac{d^2r}{dT^2} - \frac{V^2\cos^2\beta}{r} \right\}\sin\beta$$

Or

$$\frac{dv}{dT} = -\frac{GM}{r^2}\sin\beta$$

$$\text{Payload} \equiv r\frac{d^2\theta}{dT^2} + 2\frac{dr}{dT}\frac{d\theta}{dT} = \frac{1}{r}(r^2\frac{d^2\theta}{dT^2} + 2r\frac{dr}{dT}\frac{d\theta}{dT})$$

$$= \frac{1}{r}(\frac{d}{dT}(r^2\frac{d\gamma}{dT}) = \frac{1}{r}\frac{dh}{dT} = 0$$

(Payload is zero *for* Inertia and Gravitation only but bit differently!)

Amazing! Isn't it?

Which, then, of the favors of your Lord -ALLAH Kareem- the Supreme- will ye deny?

Surah Rahman, Al-Quran

Care should be exercised while reading because the same alphabets are used to denote different angles for different proofs throughout the book.

$i = \pm\sqrt{-1}$ (rotation of 90°) shows that the quantity has been rotated by 90° into the plane swept out by the line joining the Centre of rotation and the particle in Mechanical motion under a mechanical system. It also means that the quantity with which it is in relation as a coefficient lies at 90° out of the plane of mechanical motion containing the object and the center of rotation and has been rotated by 90°. So, it has two meanings, both are equally correct.

But when it is pure gravitation or gravitation alone then the couple of inertial accelerations combine to produce another condition of inertia or pure gravitation of equal areas swept in equal times.

One very important aspect to *clarify* to naïve humanity here at this juncture of Tahirian Time is that the components necessary to define the Tahirian Centrifugal/Centripetal Velocity (V″ & V′) are as below:

$$\frac{dr}{dT} = \pm \pm iV\cos\beta \quad \frac{dr}{dT} = \pm v\sin\beta$$

$$V'' = \pm v\sin\beta + \pm \pm iV\cos\beta$$

If c = a & c is also equal to b then c = a + b

Since Tahirian Rotations are real and existent, therefore the legalized values are as below, after dropping the rotation term with signs, + is anticlockwise and − is clockwise,

$$\frac{dr}{dT} = \pm V\cos\beta$$

$$V'' = \pm v\sin\beta + \pm V\cos\beta$$

$$\frac{dr}{dT} = \pm V\cos\beta \quad \& \quad h = rv\cos\beta$$

$$V'' = \pm v\sin\beta \pm V\cos\beta$$

$$V' = -v\sin\beta - V\cos\beta$$

$$V' = -(v\sin\beta + V\cos\beta) \quad \text{Tahirian Centripetal Velocity}$$

$$V'' = +v\sin\beta + V\cos\beta$$

$$V'' = +v\sin\beta + V\cos\beta \quad \text{Tahirian Centrifugal Velocity}$$

$$V'' = i^2 V'$$

Where V′ is the Tahirian Centripetal Velocity

$$V'' = v\sin\beta + V\cos\beta \quad 0 \le \beta \le +\frac{\pi}{2}$$

The notion is also true for

$$-\frac{\pi}{2} \le \beta \le 0$$

but I will not discuss it here. It can be construed but care should be exercised to keep an extra vigilant track of plus/minus signs during handling. Due to extreme dangers of mishandling, I have declared it inadmissible to keep things easy to understand for the ordinary reader. Absolute magnitude sign can also be used as well.

Ensure one fact that the combining of the two components of the velocity are true in the sense of addition and never subtraction!

$$\frac{dV''}{dT} = \left[\frac{d^2r}{dT^2} - \frac{V^2\cos^2\beta}{r}\right]$$

$$\frac{dV''}{dT} = i^2\left[\frac{V^2\cos^2\beta}{r} - \frac{d^2r}{dT^2}\right] = i^2(\text{Centripetal Acceleration})$$

$$\frac{dV''}{dT} = i^2\frac{dV'}{dT} \qquad 0 \leq \beta \leq +\frac{\pi}{2}$$

$$\frac{dV'}{dT} = \frac{V^2\cos^2\beta}{r} - \frac{d^2r}{dT^2} \qquad 0 \leq \beta \leq +\frac{\pi}{2}$$

So,

$$V' = -(v\sin\beta + V\cos\beta) \quad \text{Tahirian Centripetal Velocity}$$

After dropping the rotation of 180 degrees ($\equiv i^2$), I obtain the conclusion as below:

$$\text{T Centripetal Force} = \text{T Centrifugal Force} = \frac{V^2\cos^2\beta}{r} - \frac{d^2r}{dT^2}$$

We can show similar results if we take the derivative of the Tahirian Centripetal Velocity as given above. So there exists a third inertial force couple – Centripetal and Centrifugal Force couple in opposite directions. We see how a Centrifugal requires the existence of a Centripetal force for its explanation and vice versa. So, it's a couple at 180 degrees to each other!

$$i^2(\text{Centripetal Force}) = \text{Centrifugal Force}$$

$$i^2(\text{Centrifugal Force}) = \text{Centripetal Force}$$

Dropping Tahirian rotations, I obtain as below:

$$\text{Centripetal Force} = \text{Centrifugal Force}$$

$$\text{Centrifugal Force} = \text{Centripetal Force}$$

Inertial forces are behind all mechanical systems such as friction forces, tensions in ropes, reactions, gravitational forces, torques, compression and tension in shafts etc.

Now, let me find the power required to throw a projectile up in space, ignoring air resistance, from the surface of a planet once and forever in the Islamic History of Natural Philosophy. I will find the work done by gravity on the projectile as below and its opposite will be our goal, i.e., work done against gravity to throw it up.

$$dW = \mathbf{F}.\,\mathbf{ds}$$

$\mathbf{F}$ & $\mathbf{ds}$ are in the same direction. F is the component of gravity opposing/decelerating motion of the projectile along the direction of its motion.

$$\frac{dv}{dT} = -\frac{GM}{r^2}\sin\beta$$

$$dW = m\frac{dv}{dT}.\,ds = -m\frac{GM}{r^2}\sin\beta\, ds\cos 0^0$$

$$dW = -m\frac{GM}{r^2}\sin\beta\, ds$$

Dividing by dT on both sides, we get power required as below:

$$\frac{dW}{dT} = -m\frac{GM}{r^2}\sin\beta\frac{ds}{dT}$$

$$P = -m\frac{GM}{r^2}\sin\beta\frac{ds}{dT}$$

$$P = -m\frac{GM}{r^2}v\sin\beta$$

$$P = -m\frac{GM}{r^2}\frac{dr}{dT}$$

The above expression is now blessed with TISA Package to produce:

$$P = -m\frac{GM}{r^2}\frac{dr}{dT}$$

Where:

$$\frac{dr}{dT} = \pm \pm iV\cos\beta \quad \& \quad \frac{dr}{dT} = \pm V\sin\beta$$

If $c = a$ & c is also equal to b then $c = a + b$

$$P = -m\frac{GM}{r^2}(\pm v\sin\beta + \pm \pm iV\cos\beta)$$

According to the World Legalized System of GIVs, we obtain as below after dropping rotations:

$$P = -m\frac{GM}{r^2}(v\sin\beta + V\cos\beta)$$

$$P = -m\frac{GM}{r^2}(v\sin\beta + V\cos\beta); \qquad 0 \leq \beta \leq +\frac{\pi}{2}$$

$$P = m\frac{GM}{r^2} \times -(v\sin\beta + V\cos\beta)$$

× is a multiplication symbol and not a cross-product!

$$P = m\frac{GM}{r^2} \times -(v\sin\beta + V\cos\beta)]$$

$m\frac{GM}{r^2} \times -(v\sin\beta + V\cos\beta)$ **is the work done by gravity because it is equal to the product of Alkhazinis' gravitational force and Tahironian Centripetal Velocity (TCV) in the direction of Alkhazini's gravitational force.**

$-[m\frac{GM}{r^2} \times -(v\sin\beta + V\cos\beta)]$ **is the work done on the projectile against gravity.**

$$P = m\frac{GM}{r^2}(V\sin\beta + V\cos\beta) \quad 0 \le \beta \le +\frac{\pi}{2}$$
$$P = m\frac{GM}{r^2}(V\sin\|\beta\| + V\cos\beta) \quad -\frac{\pi}{2} \le \beta \le 0$$

This is the height of Islamic Natural Philosophy of the Tahironian Era.

Continuing from where we left off, another interesting and more intuitive way to explain is to use the Tahirian Rotatory Units (TRU) by multiplying the equation below by Tahir Rotatory Unit i.

$$\frac{dv}{dT} = [\frac{d^2r}{dT^2} - r[\frac{d\theta}{dT}]^2]\sin\beta + [-\frac{V^2\cos\beta\sin\beta}{r} + \frac{V^2\cos\beta\sin\beta}{r} - i\frac{V^2\cos^2\beta}{r} + i\frac{V^2\cos^2\beta}{r}]\cos\beta = 0$$

Or

$$\left[\frac{d^2r}{dT^2} - r\left[\frac{d\theta}{dT}\right]^2\right]\sin\beta$$
$$+\left[-\frac{V^2\cos\beta\sin\beta}{r} + \frac{V^2\cos\beta\sin\beta}{r} - i\frac{V^2\cos^2\beta}{r}\right.$$
$$\left.+ i\frac{V^2\cos^2\beta}{r}\right]\cos\beta = 0$$

So, we see that the Tahir Unit of Rotation is just meant to represent rotations for maintaining orthogonality between the three Tahirian Inertial Forces and nothing else. In other words, Tahirian Inertial Forces are *always* mutually perpendicular to each other. Writing the above equation with generalization, we obtain as below:

$$i^{n+2}\left[r\left[\frac{d\theta}{dT}\right]^2 - \frac{d^2r}{dT^2}\right]\sin\beta$$
$$+\left[i^{n+2}\frac{v^2\cos\beta\sin\beta}{r} + i^{n+0}\frac{v^2\cos\beta\sin\beta}{r}\right.$$
$$\left.+ i^{n+3}\frac{v^2\cos^2\beta}{r} + i^{n+1}\frac{v^2\cos^2\beta}{r}\right]\cos\beta = 0$$

Where:

$$i^{n+0}$$
$$i^{n+1}$$
$$i^{n+2}$$
$$i^{n+3}$$

are just Tahir Units *of* Rotations

$$i^0 = +1 \quad 0^\circ$$
$$i^1 = +i \quad 90^\circ$$
$$i^2 = -1 \quad 180^\circ$$
$$i^3 = -i \quad 270^\circ$$
$$i^4 = +1 \quad 360^\circ$$
$$\cdots$$
$$i^n = e^{\frac{in\pi}{2}} \quad 90 \times n^\circ$$

Where $n = 0, 1, 2, 3, 4, 5, \ldots\ldots\ldots \infty$

Speaking on Tahir Scale Consciousness (TSC), Tahirian Cosmos is NOT empty even when there is no matter in it and a point is moving in a straight line at uniform speed and a rotational speed around its own axis!

So, basically three different types of inertial acceleration couples are acting in Tahirian Inertia as in the figure below:

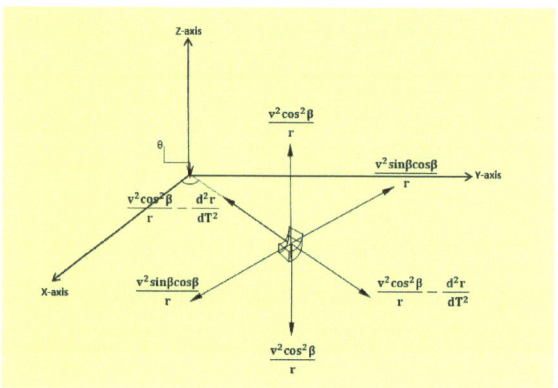

Figure 5. *An element of mass* dm *in motion, in x-y plane, at constant velocity in a straight line in inertia or matter-free space under the cancelling effect of Tahirian Inertial Acceleration Couples.*

The gravitational force also establishes a coupled relationship between two particles of matter, e.g., binary stars, as is done by all three mutually perpendicular Tahirian Inertial Force couples.

When directions are shown with arrows then signs are unimportant.

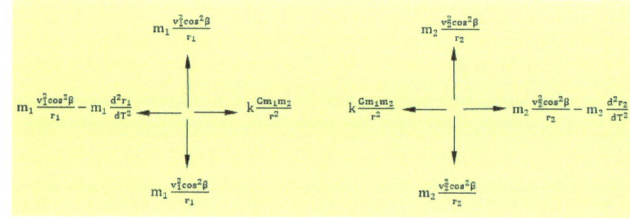

Figure 5a. Two Body Motion. *Two masses moving at varying velocities under binary star motion (gravitational motion) in precessing elliptical orbits around a fixed TCR.* $\beta_1 = \beta_2 = \beta$ & $T_1 = T_2 = T$. $v_1 \ne v_2$.

The following holds true for figure 5 after multiplying by mass m.

$$\sum F_{radial} = 0$$
$$\sum F_{orthogonal} = 0$$

$$\sum F_{z\,axis} = 0$$

Or if motion is considered by any massive object with a net acceleration in the direction of motion, then the equation

$$\sum F_S = m\frac{dv}{dT} = m\frac{d^2S}{dT^2} = ma$$

Centripetal/centrifugal forces are like any other mechanical forces i.e., gravity, friction, compression, tension etc. Centripetal and centrifugal forces can both be mechanical/physical and Inertial forces. They make couples only in opposition as below:

Inertial Centripetal force = Mechanical Centrifugal force

Inertial Centrifugal force = Mechanical Centripetal force

The above first condition is applicable to Gyroscope when rotating at an angle below the horizontal.

I will form Tahirian Triads *of* Inertial Accelerations and Forces (TTIF) as below:

$$m\frac{v^2\cos^2\beta}{r} - m\frac{d^2r}{dT^2}, \quad m\frac{v^2\sin\beta\cos\beta}{r} \quad \& \quad m\frac{v^2\cos^2\beta}{r} \qquad \text{TTIF-1}$$

$$m\frac{v^2\cos^2\beta}{r} - m\frac{d^2r}{dT^2}, \quad m\frac{dr}{dT}\frac{d\theta}{dT} \quad \& \quad m\frac{v^2\cos^2\beta}{r} \qquad \text{TTIF-2}$$

$$m\frac{v^2\cos^2\beta}{r} - m\frac{d^2r}{dT^2}, \quad mr\frac{d^2\theta}{dT^2} \quad \& \quad m\frac{v^2\cos^2\beta}{r} \qquad \text{TTIF-3}$$

$$m\frac{d^2r}{dT^2} \qquad \text{or} \qquad m\frac{d^2S}{dT^2}$$

Now, let me apply the Triads to solve the problems. Suppose a solid cylindrical piece of mass m and length L is rolling down an incline at an angle of θ. Calculate the acceleration of the roller? Given the coefficient of friction is μ.

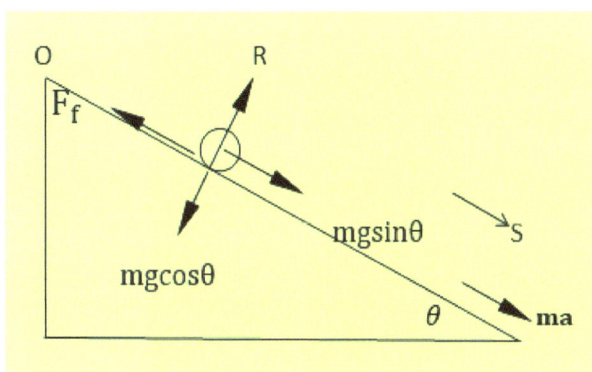

Figure 6. *A solid cylindrical piece rolling down on an incline.*

$$\sum F_S = m\frac{dv}{dT} = m\frac{d^2S}{dT^2} = ma$$

Applying the above equation, we obtain as below:

$$-F_f + mg\sin\theta = ma$$

Triad will be used to find the torque developed by frictional force in the above rolling cylinder as below:

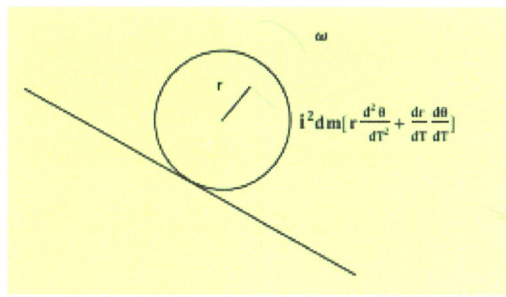

Figure 7. *Force acting on the differential element dm of the cross section of the solid cylindrical piece rolling down on the incline.*

θ can be considered as the angle between the radial line and the vertical line, passing through the center, instead of the horizontal line passing through the same center. $\frac{dr}{dT} = 0$ because of which the second term is zero. A *fixed* differential element of mass dm is considered which is not shown in the above figure but it lies at the corner where the right angle is formed.

$$d\tau = dF \times r = i^2 dm\, r\frac{d^2\theta}{dT^2} \times r$$

$$d\tau = i^2 r^2 dm\frac{d^2\theta}{dT^2}$$

$$\frac{d^2\theta}{dT^2} = \alpha \quad \text{constant angular acceleration}$$

$$d\tau = i^2 r^2 dm\alpha$$

$$\int_0^\tau d\tau = i^2\alpha\int_0^R r^2 dm$$

$$\tau_{resistive} = i^2 I\alpha$$

The above is the resistive torque, the applied torque is:

$$\tau_{applied} = I\alpha$$

Furthermore, the applied torque is due to frictional force as given below:

$$\tau_{applied} = I\alpha = F_f \times R$$

$$F_f = \frac{I\alpha}{R} = \frac{1}{2}mR\alpha$$

$$F_f = \frac{1}{2}mR\alpha = \frac{1}{2}ma$$

$$-F_f + mg\sin\theta = ma$$

$$-F_f + mg\sin\theta = ma$$

$$-\frac{1}{2}ma + mg\sin\theta = ma$$

$$a = \frac{2}{3}g\sin\theta$$

In the above problem we have utilized two Tahirian Triads to solve it.

Now let me find the tension in an inextensible string rotating a ball of mass m in horizontal circle for the first time in World History.

In the previous example I have used one part of the Monad

$$m\frac{v^2\cos^2\beta}{r} - m\frac{d^2r}{dT^2}$$

but in this example, I will use the other part of the Tahirian Monad as below:

$$m\frac{v^2}{r} \; ; \quad \beta = 0$$

Basically, I am using the TTIF-1 from the list of Tahirian Triads.

I will solve this problem with the condition, and not the assumption $\frac{d\theta}{dT} = 0$. In other words, at constant speed in the horizontal circle. The second term in the Tahirian Centrifugal Force is $\frac{d^2r}{dT^2}$ and is equal to zero because the string is assumed inextensible. What is happening here is that two mechanical forces or their rectangular components are competing with the Forces *of* Tahir (Inertial Forces). X direction is in the direction of the line of action of Tension and the y direction is perpendicular to it.

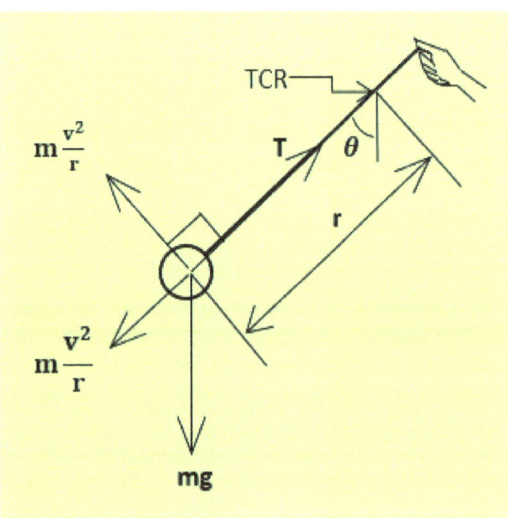

Figure 8. *Forces acting on an inextensible string rotating a ball of mass m in a horizontal circle.*

So, from the above figure, summation of forces in the x direction gives Tahir:

$$\sum F_x = 0$$

$$T - m\frac{v^2}{r} - mg\cos\theta = 0$$

Similarly, summation of forces in the y direction gives Tahir another equation as below:

$$\sum F_y = 0$$

$$m\frac{v^2}{r} - mg\sin\theta = 0$$

$$T - m\frac{v^2}{r} = mg\cos\theta$$

$$m\frac{v^2}{r} = mg\sin\theta$$

$$T = m\frac{v^2}{r} + mg\cos\theta$$

$$T = m\frac{v^2}{r} + \sqrt{m^2g^2 - (m\frac{v^2}{r})^2} \quad \text{for } g \geq \frac{v^2}{r}$$

$$T = m\frac{v^2}{r} + m\sqrt{g^2 - \frac{v^4}{r^2}} \quad \text{for } g \geq \frac{v^2}{r}$$

$$T = m[\frac{v^2}{r} + \sqrt{g^2 - \frac{v^4}{r^2}}] \quad \text{for } g \geq \frac{v^2}{r}$$

For $\frac{v^2}{r} \geq g$, we need to proceed cautiously as below because the angle θ becomes $90°$. A portion of the vertical force which is not needed anymore in the balancing of the forces in the vertical direction is now rotated $90°$ into the plane of the rotating string rotating horizontally or in the direction of the force of Tension. *Amazing, isn't it?*

$$T = m\frac{v^2}{r} + \sqrt{m^2g^2 - m^2\frac{v^4}{r^2}}$$

$$T = m\frac{v^2}{r} + \sqrt{-m^2\frac{v^4}{r^2} + m^2g^2}$$

$$T = m\frac{v^2}{r} + \sqrt{-m^2\frac{v^4}{r^2}(1 - \frac{g^2R^2}{v^4})}$$

$$T = m\frac{v^2}{r} + \pm im\frac{v^2}{r}\sqrt{1 - \frac{g^2R^2}{v^4}}$$

$$\text{Where } i = \sqrt{-1}$$

$$T = m\frac{v^2}{r} + \pm im\frac{v^2}{r}\beta$$

$$\text{Where } \beta = \sqrt{1 - \frac{g^2R^2}{v^4}} \quad \& \quad 0 \leq \beta < 1$$

As per the laws of the World Legalized System of Values of the Golden Islamic Values (GIV's)

$$T = m\frac{v^2}{r}[1+\beta]$$

The positive and negative signs are associated *only* with the 90° rotations on either side of the vertical line!!! When we drop the rotation term $\pm i$ we must drop both positive and negative signs associated with it leaving the original algebraic positive sign there!!!

So, I have

$$T = m\frac{v^2}{r}[1+\beta] \quad \text{Where } \frac{v^2}{r} \geq g \quad \theta = 90°$$

The new Centrifugal force on the left is now as below:

$$m\frac{v^2}{r}[1+\beta]$$

Which is the same as the Tension in the string!!!

Because rotation of a portion of the vertical force happens on both sides (two signs plus and minus with i) of the vertical line at

$$\theta = 90° \quad \text{Where } \frac{v^2}{r} \geq g \quad !!!$$

So, the Tension in the string is given as below:

$$T = m[\frac{v^2}{r} + \sqrt{g^2 - \frac{v^4}{r^2}}] \quad \text{for} \quad g \geq \frac{v^2}{r} \ \& \ 0 \leq \theta < 90°$$

$$T = m\frac{v^2}{r}[1+\beta] \quad \text{Where } \frac{v^2}{r} \geq g \quad \theta = 90°$$

Amazing, isn't it?

Frankly speaking, the component Vcosβ, with the assistance of Tahir Intellectual & Spiritual Aid (TISA), breaks up into two components as below

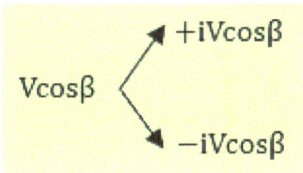

Figure 8a. Symmetry in the making

Now let Fayaz Tahir find the maximum Tension in the string as follows:

$$T = m\frac{v^2}{r} + \sqrt{m^2g^2 - m^2\frac{v^4}{r^2}}$$

$$\frac{dT}{dv} = 0$$

$$\frac{v^2}{r} = \frac{g}{\sqrt{2}}$$

Differentiation is left as an exercise for the naives (gololio polio, noton, anstan, maglell, ooler, nother etc.) of my Time, before and after.

Where are your gololio polios and their naive thoughts, where are your copernicoons and their revolutions? Where are your jawan kapils and their planetary laws?

Substituting the above value back in the Tahirian Equation below, I have

$$m\frac{v^2}{r} = mg\sin\theta$$

$$\frac{v^2}{r} = g\sin\theta$$

$$\sin\theta = \frac{v^2}{gr} = \frac{1}{\sqrt{2}}$$

θ = 45° (for maximum Tension)

The ill-educated pagans of Copenhagen, neel boar etc., couldn't feel this maximum Tension because their brains had become numb for the last just 1000 years.

The tension is given as below:

$$T = m\frac{v^2}{r} + \sqrt{m^2g^2 - m^2g^2\sin^2\theta} \quad \text{for } g \geq \frac{v^2}{r}$$

Inserting the value, I obtain

$$T_{max} = m\frac{v^2}{r} + \frac{1}{\sqrt{2}}mg$$

$$T_{max} = \frac{1}{\sqrt{2}}mg + \frac{1}{\sqrt{2}}mg$$

$$T_{max} = \sqrt{2}mg \quad \text{for } g \geq \frac{v^2}{r}$$

We also see that the minimum Tension occurs at $\theta = 90°$ and $\theta = 0°$ and is equal to

$$T_{min} = mg \quad \text{for } g \geq \frac{v^2}{r}$$

Now let Tahir solve the mysterious of the most mysterious problems, i.e., the Gyroscope problem as below once and forever in the History of Islamic Physics.

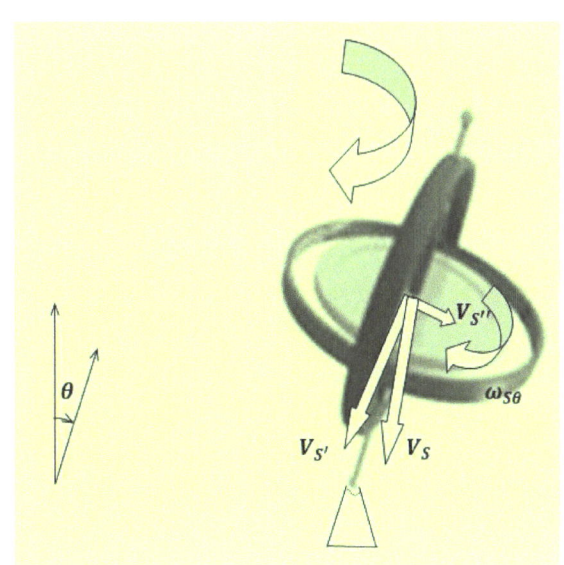

Figure 9. *Resolving the resultant velocity of a Gyroscope into its rectangular components.*

The best way to approach this problem is to break up the resultant velocity of the gyroscope into its rectangular components. *Basically, I have broken one motion* V_S *into two motions* $V_{S'}$ *and* $V_{S''}$ for the Gyroscope falling or rising. When the Gyroscope is inclined at an angle, it is either falling or rising. The Gyroscope that I have analyzed is a particular one shown here in figures. Different types of Gyroscopes have different shapes and hence different analysis. In the above figure, the two rectangular components of the resultant velocity of the gyroscope and the resultant itself are coplanar and are in the plane of the gyroscope rotor. So, the first Tahironian Equation that can be made by applying Theorem de Alkashi is as below:

$$V_S^2 = V_{S'}^2 + V_{S''}^2$$

Let the spinning axis of the rotor represents x axis and the axis perpendicular to it, in the plane of the rotor and in the direction of one of the forces as represented below, be y axis.

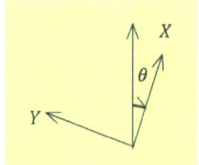

Figure 10. *Reference axes.*

So, from the above figure, summation of forces in the y direction gives as below. The gyroscope starts rising with high rotor launch speed $\omega_{S\theta'}$, not shown in the figure. The direction of $V_{S''}$ should be considered in accordance with the Gyroscope going up or down.

$$\sum F_{S''} = M\frac{dV_{S''}}{dT} = M\frac{d^2S''}{dT^2} = Ma''$$

$$M\frac{V_{S'}^2}{R} - Mg\sin\theta = M\frac{dV_{S''}}{dT}$$

$$M\frac{V_{S'}^2}{R} - Mg\sin\theta = MR\frac{d}{dT}\left[\frac{d\theta}{dT}\right]$$

$$M\frac{V_{S'}^2}{R} - Mg\sin\theta = MR\frac{d}{d\theta}\left[\frac{d\theta}{dT}\right]\frac{d\theta}{dT}$$

$$M\kappa^2R - Mg\sin\theta = MR\omega\frac{d\omega}{d\theta}$$

$$\kappa^2R - g\sin\theta = R\omega\frac{d\omega}{d\theta}$$

The negative sign below is due to the gyroscope rising and causing ω to be negative as below, but it doesn't make any difference because the negative sign cancels out.

$$\pm\omega = \frac{d\theta}{dT}$$

Whereas, in the diagram the gyroscope is shown coming down with $V_{S''}$, in order, to avoid the force vectors merging on each other.

Where $\kappa = \frac{d\Omega}{dT}$ and Ω is the angle traversed by the gyroscope rotor along the *slant surface* of the cone, that it generates while under Tahironian Motion, and should not be confused with the actual spirally slant surface generated by the gyroscope while rising and coming down depending upon the speed V_S of the rotor, not shown in the Tahironian Diagrams.

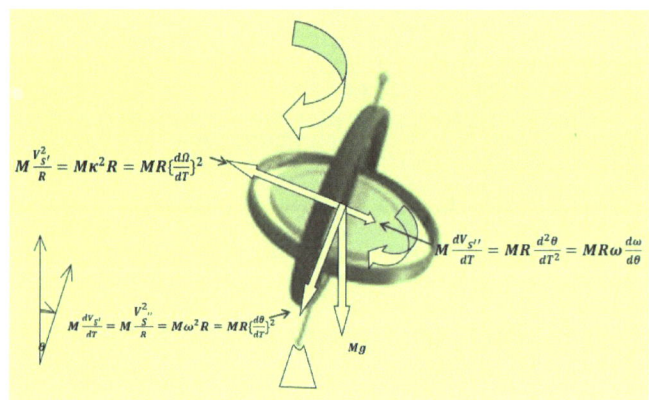

Figure 11. *Analysis of Forces on the Gyroscope in the y direction.*

The lengths of the velocity and force vectors should not be considered representative of their respective magnitudes because they are not drawn to scale. The three forces above are acting in the plane of the rotor except gravity which is acting vertically downwards.

So, again, from the above figure, summation of forces in the z direction, perpendicular to the plane containing x & y axes and not shown in figures, gives as below. The gyroscope is rising with high rotor speed ω_{rs}, not shown in the figure. The figure shows the downward motion only.

$$\sum F_{S'} = M\frac{dV_{S'}}{dT} = M\frac{d^2S'}{dT^2} = Ma'$$

$$M\frac{V_{S''}^2}{R} - Mg\cos90^o = M\frac{dV_{S'}}{dT}$$

$$M\omega^2R - Mg\cos90^o = M\frac{dV_{S'}}{dT}$$

$$M\omega^2R = M\frac{dV_{S'}}{dT}$$

$$M\omega^2R = M\frac{dV_{S'}}{dT} = MR\frac{d\kappa}{dT}$$

$$\omega^2 = \frac{d\kappa}{dT}$$

$$\omega^2 = \frac{d\kappa}{d\theta}\frac{d\theta}{dT}$$

$$\omega^2 = \pm\omega\frac{d\kappa}{d\theta}$$

$$\omega = \pm\frac{d\kappa}{d\theta}$$

In the case of Gyroscope rising, we have as below:

$$\omega = -\frac{d\kappa}{d\theta}$$

Now let me write the third equation as below:

Summing up forces in the vertical direction, we obtain as below:

$$V - mg + \frac{mV_s^2}{R}(\sin\theta + \cos\theta) + \left[\int_0^r \frac{(dm)v^2}{r}\right]\cos\theta = 0$$

$$V - mg + \frac{mV_s^2}{R}(\sin\theta + \cos\theta) + \left[\int_0^r \frac{dm\,\omega_{rs}^2 r^2}{r}\right]\cos\theta = 0$$

$$V - mg + \frac{mV_s^2}{R}(\sin\theta + \cos\theta) + \left[\int_0^r dm\,\omega_{rs}^2 r\right]\cos\theta = 0$$

$$V - mg + \frac{mV_s^2}{R}(\sin\theta + \cos\theta) + \left[\int_0^r \omega_{rs}^2 r\,dm\right]\cos\theta = 0$$

Vertical component V of the reaction at the base of the Gyroscope can be measured by placing the Gyroscope on a delicate weighing instrument fitted with a graphing screen that can measure and note the vertical component of the reaction at different values of the angle θ. The average value of V can then be taken of all values thus obtained.

The rotor spin speed (rs) square value can be easily taken out of the integral sign *unaveraged* because it is not dependent upon the materials' geometric properties even when the density of the material is considered varying throughout the mass distribution. V is the reaction at the base of the Gyroscope.

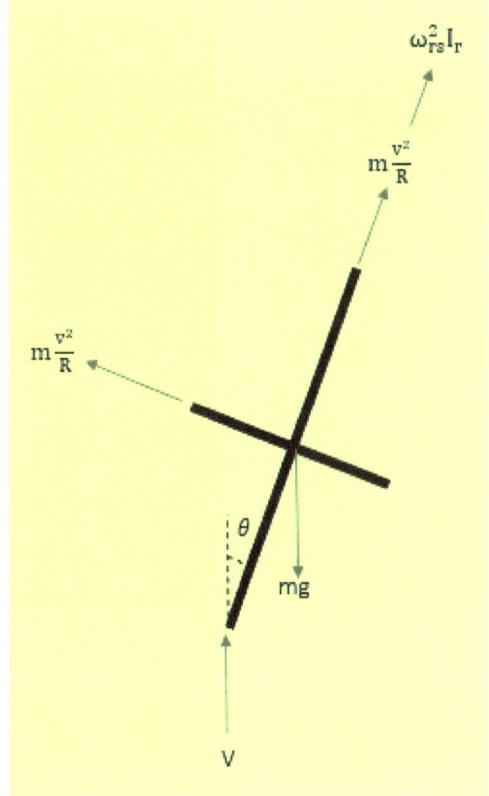

Figure 12. *Analysis of Forces on the Gyroscope in the vertical direction for* $\theta \leq \frac{\pi}{2}$

$$V - mg + \frac{mV_s^2}{R}(\sin\theta + \cos\theta) + \omega_{rs}^2\left[\int_0^r r\,dm\right]\cos\theta = 0$$

$$V - mg + \frac{mV_s^2}{R}(\sin\theta + \cos\theta) + \omega_{rs}^2 I_r\cos\theta = 0$$

$$I_r = \int_0^r r\,dm$$

So, we have the two equations as below:

$$\kappa^2 R - g\sin\theta = R\omega\frac{d\omega}{d\theta}$$

$$V - mg + \frac{mV_s^2}{R}(\sin\theta + \cos\theta) + \omega_{rs}^2 I_r\cos\theta = 0$$

$$V - mg + \frac{mV_s^2}{R}(\sin\theta + \cos\theta) + \omega_{rs}^2 I_r\cos\theta = 0$$

The above equation can also be written as

$$V - mg + \frac{m(v_s^2 + v_{s''}^2)}{R}(\sin\theta + \cos\theta) + \omega_{rs}^2 I_r\cos\theta = 0$$

$$V - mg + mR(\kappa^2 + \omega^2)(\sin\theta + \cos\theta) + \omega_{rs}^2 I_r\cos\theta = 0$$

$$V - mg + mR\left(\frac{g\sin\theta}{R} + \omega\frac{d\omega}{d\theta} + \omega^2\right)(\sin\theta + \cos\theta) + \omega_{rs}^2 I_r\cos\theta = 0$$

$$mR\left(\frac{g\sin\theta}{R} + \omega\frac{d\omega}{d\theta} + \omega^2\right)(\sin\theta + \cos\theta) = mg - V - \omega_{rs}^2 I_r\cos\theta$$

$$\omega\frac{d\omega}{d\theta} + \omega^2 = \frac{mg - V - \omega_{rs}^2 I_r\cos\theta}{mR(\sin\theta + \cos\theta)} - \frac{g\sin\theta}{R}$$

$$\frac{d\omega}{d\theta} + \omega = [\frac{mg - V - \omega_{rs}^2 I_r\cos\theta}{mR(\sin\theta + \cos\theta)} - \frac{g\sin\theta}{R}]\omega^{-1}$$

The above equation is of the form:

$$\frac{dy}{dx} + P(x)y = Q(x)y^n$$

And its solution is:

$$\sigma\mu = \int e^{\int(1-n)P(x)dx}(1-n)Q(x)dx$$

Where $\sigma(x) = y^{1-n}$ **&** $\mu(x) = e^{\int(1-n)P(x)dx}$

$P(\theta) = 1$, $Q(\theta) = \frac{mg-V-\omega_{rs}^2 I_r\cos\theta}{mR(\sin\theta+\cos\theta)} - \frac{g\sin\theta}{R}$ & $n = -1$

$$\omega^2 e^{2\theta} = \int 2e^{2\theta}\left(\frac{mg - V - \omega_{rs}^2 I_r\cos\theta}{mR(\sin\theta + \cos\theta)} - \frac{g\sin\theta}{R}\right)d\theta$$

$$\omega^2 e^{2\theta} = \int_\theta^0 2e^{2\theta}\left(\frac{mg - V - \omega_{rs}^2 I_r\cos\theta}{mR(\sin\theta + \cos\theta)} - \frac{g\sin\theta}{R}\right)\theta$$

ω_{rs}^2 cannot be taken out of the integral sign unaveraged because it is dependent upon T & θ. Hence, we will take the averaged value and represent it with a bar on top of it. Similar is the case with the vertical component of reaction V at the base of the Gyroscope. V is dependent upon θ.

$$\omega^2 e^{2\theta} = 2(\frac{g}{R} - \frac{\bar{V}}{mR})\int\frac{e^{2\theta}}{\sin\theta + \cos\theta}d\theta - 2\frac{\overline{\omega_{rs}^2}I_r}{mR}\int\frac{e^{2\theta}\cos\theta}{\sin\theta + \cos\theta}d\theta - 2\frac{g}{R}\int e^{2\theta}\sin\theta d\theta$$

So basically, we need to solve the two integrals above. The last integral can be easily solved. One way to solve the definite integrals is to use the Tahir Series expansion of functions as below:

$$\int_{\theta_1}^{\theta_2} f(\theta)d\theta = \left|\theta f(\theta) - \frac{\theta^2}{2!}f^1(\theta) + \frac{\theta^3}{3!}f^2(\theta) - \frac{\theta^4}{4!}f^3(\theta) + \cdots\right|_{\theta_1}^{\theta_2}$$

When one function is exponential then another way, which is simpler than the above, is the following:

$$\int_{\theta_1}^{\theta_2} e^\theta f(\theta)d\theta = \left|e^\theta\{f(\theta) - f^1(\theta) + f^2(\theta) - f^3(\theta) + \cdots\}\right|_{\theta_1}^{\theta_2}$$

Another method is to substitute the Tahir Series expansions of sine, cosine and the exponential functions and reduce the resulting expression to the desired power of the angle by Long Division. One more interesting method of solving definite integrals is the mean value method which I never saw in any naïve book written by northerners, e.g. russians, americans etc. I think its convergence is quicker than many methods. I might discuss that method somewhere else in my book if time permits.

Since we need to find the value of **κ** as well, the best method is to use the Tahir Series Expansions of sine, cosine and the exponentials and find an expression in terms of the angle by using long division method in the end. Which we can then integrate to find **κ** in terms of the angle.

$$\omega_{rs} \cong -\frac{\omega_{rs_{\theta_L}}}{T_t}T + \omega_{rs_{\theta_L}}$$

$$\omega_{rs_{\theta_0}} \cong -\frac{\omega_{rs_{\theta_L}}}{T_t}T_0 + \omega_{rs_{\theta_L}}$$

$$\overline{\omega_{rs}^2} \cong \frac{\omega_{rs_{\theta_L}}^2 + \omega_{rs_{\theta_0}}^2}{2}$$

$$\omega_{rs_{\theta_c}} \cong -\frac{\omega_{rs_{\theta_L}}}{T_t}T_{\theta_c} + \omega_{rs_{\theta_L}}$$

θ_L is the angle of launch at which the high precision Gyroscope has been positioned in its orbit before it starts going up at sufficiently high rotor spin speed and reaches the angle zero θ_0 while vertically straight. T_0 is the time from the launch to when the Gyroscope gets vertical at angle zero. T_t is the total time from the launch to when the Gyroscope rotor spin comes to a complete stop long after the Gyroscopic motion collapses. Basically, the rotor speed is dependent upon one variable, Time. Since gravitational field is irrotational, hence when we give rotations to an object and let it fall under gravity then gravity can't disturb its rotations but can keep it in an orbit following Tahirian Trajectories of the projectiles or of Gyroscope or any other Tahirian Phenomenon in Tahirian Cosmology. The main player of the whole game is always gravity *but at the center of mass only.* $\omega_{rs_{\theta_L}}$ should be known in the beginning.

A body continues in its state of uniform (constant angular speed) rotational/spinning motion about a fixed axis unless compelled by some internal or external force/torque to act otherwise.

$$\omega_{rs} = \omega_{rs}(T)$$

$$\omega(\theta) = -\frac{d\kappa}{d\theta}$$

$$\kappa = -\int_\theta^0 \omega(\theta)d\theta$$

$$\kappa = -\int_\theta^0 \omega(\theta)d\theta = -\int_\theta^0 \omega d\theta$$

$$\kappa = -\int_\theta^0 \omega d\theta = -\left|\theta\omega - \frac{\theta^2}{2!}\omega^1(\theta) + \frac{\theta^3}{3!}\omega^2(\theta) - \frac{\theta^4}{4!}\omega^3(\theta) + \cdots\right|_\theta^0$$

Now let me run a similar analysis of the same Gyroscope when it has started falling after it has stood upright for some time. The terms are similar but with opposite signs this time.

So, we have the two equations as below:

$$-\kappa^2 R + g\sin\theta = R\omega\frac{d\omega}{d\theta}$$

$$V - mg + \frac{mV_s^2}{R}(\sin\theta + \cos\theta) + \omega_{rs}^2 I_r\cos\theta = 0$$

$$V - mg + \frac{m\left(v_{s'}^2 + v_{s''}^2\right)}{R}(\sin\theta + \cos\theta) + \omega_{rs}^2 I_r \cos\theta = 0$$

$$V - mg + mR(\kappa^2 + \omega^2)(\sin\theta + \cos\theta) + \omega_{rs}^2 I_r \cos\theta = 0$$

$$V - mg + mR\left(\frac{g\sin\theta}{R} - \omega\frac{d\omega}{d\theta} + \omega^2\right)(\sin\theta + \cos\theta) + \omega_{rs}^2 I_r \cos\theta = 0$$

$$mR\left(\frac{g\sin\theta}{R} - \omega\frac{d\omega}{d\theta} + \omega^2\right)(\sin\theta + \cos\theta) = mg - V - \omega_{rs}^2 I_r \cos\theta$$

$$-\omega\frac{d\omega}{d\theta} + \omega^2 = \frac{mg - V - \omega_{rs}^2 I_r \cos\theta}{mR(\sin\theta + \cos\theta)} - \frac{g\sin\theta}{R}$$

$$\frac{d\omega}{d\theta} - \omega = \left[\frac{V + \omega_{rs}^2 I_r \cos\theta - mg}{mR(\sin\theta + \cos\theta)} + \frac{g\sin\theta}{R}\right]\omega^{-1}$$

The above equation is of the form:

$$\frac{dy}{dx} + P(x)y = Q(x)y^n$$

And its solution is:

$$\sigma\mu = \int e^{\int(1-n)P(x)dx}(1-n)Q(x)dx$$

Where $\sigma(x) = y^{1-n}$ & $\mu(x) = e^{\int(1-n)P(x)dx}$

$$P(\theta) = -1, \; Q(\theta) = \frac{V + \omega_{rs}^2 I_r \cos\theta - mg}{mR(\sin\theta + \cos\theta)} + \frac{g\sin\theta}{R} \; \& \; n = -1$$

$$\omega^2 e^{-2\theta} = \int_0^\theta 2e^{-2\theta}\left(\frac{V + \omega_{rs}^2 I_r \cos\theta - mg}{mR(\sin\theta + \cos\theta)} + \frac{g\sin\theta}{R}\right)d\theta$$

$$\omega^2 e^{-2\theta} = 2\left(\frac{\bar{V}}{mR} - \frac{g}{R}\right)\int \frac{e^{-2\theta}}{\sin\theta + \cos\theta}d\theta + 2\frac{\overline{\omega_{rs}^2 I_r}}{mR}\int\frac{e^{-2\theta}\cos\theta}{\sin\theta + \cos\theta}d\theta + 2\frac{g}{R}\int e^{-2\theta}\sin\theta d\theta$$

$$\omega(\theta) = +\frac{d\kappa}{d\theta}$$

$$\kappa = \int_0^\theta \omega(\theta)d\theta$$

$$\kappa = \int_0^\theta \omega(\theta)d\theta = \int_0^\theta \omega d\theta$$

$$\kappa = \int_0^\theta \omega d\theta = \left|\theta\omega - \frac{\theta^2}{2!}\omega^1(\theta) + \frac{\theta^3}{3!}\omega^2(\theta) - \frac{\theta^4}{4!}\omega^3(\theta) + \cdots\right|_\theta^0$$

Once ω is found by applying Tahir Series expansion of functions on sine, cosine and exponential followed by long division, κ can then be found by integrating the final expression.

The above analysis corresponds to

$$0 \le \theta \le \frac{\pi}{2}$$

&

$$0 \le \theta_c \le \frac{\pi}{2}$$

And the fact that sufficiently high spin speed is given to the rotor of the high precision Gyroscope that it goes up and stays vertical for some time and then starts falling till it eventually collapses at an angle θ_c.

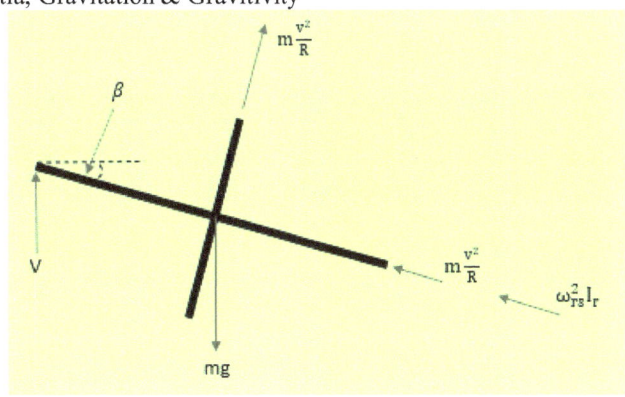

Figure 13. *Analysis of Forces on the Gyroscope in the vertical direction for* $\theta \ge \frac{\pi}{2}$

Now let me run the analysis for the following case while the Gyroscope is coming down.

$$\theta_c > \theta > \frac{\pi}{2}$$

$$-\kappa^2 R + g\sin\theta = R\omega\frac{d\omega}{d\theta}$$

$$V - mg + \frac{mV_s^2}{R}(\sin\beta + \cos\beta) + \omega_{rs}^2 I_r \sin\beta = 0$$

Where $\theta = \frac{\pi}{2} + \beta$

In the above scenario

$$\omega = \frac{d\theta}{dT} = \frac{d\beta}{dT}$$

So, the above equations become:

$$-\kappa^2 R + g\cos\beta = R\omega\frac{d\omega}{d\beta}$$

$$V - mg + \frac{mV_s^2}{R}(\sin\beta + \cos\beta) + \omega_{rs}^2 I_r \sin\beta = 0$$

$$\omega^2 e^{-2\beta} = \int_0^\beta 2e^{-2\beta}\left(\frac{V + \omega_{rs}^2 I_r \sin\beta - mg}{mR(\sin\beta + \cos\beta)} + \frac{g\cos\beta}{R}\right)d\beta$$

$$\omega^2 e^{-2\beta} = 2\left(\frac{\bar{V}}{mR} - \frac{g}{R}\right)\int\frac{e^{-2\beta}}{\sin\beta + \cos\beta}d\beta + 2\frac{\overline{\omega_{rs}^2 I_r}}{mR}\int\frac{e^{-2\beta}\sin\beta}{\sin\beta + \cos\beta}d\beta + 2\frac{g}{R}\int e^{-2\beta}\cos\beta d\beta$$

Where $\theta_c - \frac{\pi}{2} > \beta > 0$

This was the story of the Gyroscope!

Naïve Humanity! lend me your ears. There are basically two types of Tahironian Centripetal Forces. One Centripetal force is inertial and the second one is Mechanical or Physical. Similarly, there are two types of Tahironian Centrifugal forces. One is inertial and the other is Mechanical or Physical. They cross each other to make couples as below:

Inertial Centripetal force $=$ Mechanical Centrifugal force

Inertial Centrifugal force $=$ Mechanical Centripetal force

Just a reminder! Nothing serious.

Now since gravity is the main player of the whole game of Islamic Physics. When gravity produces a mechanical centripetal force like the tension in the string with a mass attached to its end and moving in a horizontal circle as discussed earlier then the inertial centripetal force is replaced by the Tension in the string which is counteracted by inertial centrifugal force of the Inertial Couple. Similarly, when gravity produces a mechanical centrifugal force i.e., tension force like the one in the gyroscope rotor axis shaft at angle β in the above last scenario of the Gyroscopic motion then the inertial centripetal force counteracts the mechanical centrifugal force. The mechanical centrifugal force replaces the inertial Centrifugal force. Whereas at angle θ the same gravity produces a mechanical centripetal force, i.e., a compressive force in the shaft of the rotor axis then the inertial centripetal force acts as a counteractive force i.e., inertial centrifugal force. *Amazing, isn't it?*

There is a tug of war between the mechanical/physical force of gravity, and other mechanical/physical forces induced due to the same gravitational force, and the Tahironian Inertial Forces in dynamical motion under Tahironian Torques. But when the bodies are at rest or under static motion then gravity induces physical/mechanical forces like tension/compression or reactions at supports and Tahironian Bending moments because v = 0. Amazing, isn't it? At rest it is Tahironian Civil Engineering and in motion it is Tahironian Celestial Mechanics.

Care should be exercised to add the constant of integration right after we have integrated and right before initiating anymore algebraic manipulation. Otherwise, the answer would not be correct.

The spin of the gimbal and the frame of the gyroscope is due to the frictional-forces and is a manufacturing issue. A friction-induced mathematical model can be developed to find the approximate speed of spin of the gimbal and frame of the gyroscope, or it can be measured manually by using a stopwatch because the speed is not as high as the speed of the rotor of the gyroscope.

The rotor-axis combination induces frictional forces to the Gimbal-Frame-Structure (GFS) axis combination due to which the GFS combination rotates in the same direction as the rotor but with a small angular speed.

There are basically two axes of rotations separated from each other by frictional forces of contact due to manufacturing technique which differs from manufacturer to manufacturer. The more the frictional-forces the lesser the speed of the rotor-axis combination and the more the speed of the GFS axis combination. Greater frictional force reduces the speed of rotor-axis combination and increases the GFS axis combination speed. Gyroscopes are symmetrical and can be used upside down with negligible loss of generality.

Men have heroes when they are not heroes themselves.

So, who is your hero my dear reader?

Let me prove the above "Tahir Principle of Instantaneity" now in a different manner because all sorts of keenii are going to teach my theory in their classrooms all over the globe and I would take care of my Tahirians of all cultures.

Using the most graceful theorem of entire mathematics i.e., the theorem of the Muslim Father of Plane Trigonometry Muhammed Ghyiasuddin Jamshed Al-Kashi namely Theorem de Al-kashi, we can write regarding the previous figure as below:

$$\overline{CD}^2 = \overline{OC}^2 + \overline{OD}^2 - 2 * \overline{OC} * \overline{OD} * \cos\theta$$

$$v^2T^2 = (R + h_1)^2 + (R + h_2 + h_3)^2 - 2(R + h_1)(R + h_2 + h_3)\cos\theta$$

Differentiating the above equation with respect to T we obtain

$$v^2T = (R + h_2 + h_3)\left(\frac{dh_2}{dT} + \frac{dh_3}{dT}\right) - (R + h_1)\{(R + h_2 + h_3)\left(-\sin\theta\frac{d\theta}{dT}\right) + \cos\theta\left(\frac{dh_2}{dT} + \frac{dh_3}{dT}\right)\}$$

$R + h_1$ must be considered constant during the process of differentiation as before and this is because we need to find the limiting values when $\theta \to 0$.

Differentiating the above equation once again with respect to T we obtain:

$$v^2 = \left(\frac{dh_2}{dT} + \frac{dh_3}{dT}\right)^2 + (R + h_2 + h_3)\left(\frac{d^2h_2}{dT^2} + \frac{d^2h_3}{dT^2}\right) - (R + h_1)[\left(\frac{dh_2}{dT} + \frac{dh_3}{dT}\right)\left(-\sin\theta\frac{d\theta}{dT}\right)$$

$$-(R + h_2 + h_3)\cos\theta\left(\frac{d\theta}{dT}\right)^2 - (R + h_2 + h_3)\sin\theta\frac{d^2\theta}{dT^2} + \cos\theta\left(\frac{d^2h_2}{dT^2} + \frac{d^2h_3}{dT^2}\right)$$

$$-\sin\theta\frac{d\theta}{dT}\left(\frac{dh_2}{dT} + \frac{dh_3}{dT}\right)]$$

Now putting the limits i.e., when T tends to 0, h_2 tends to h_1 & h_3 tends to 0 as before and, also, noting the fact that

$$r = R + h_1$$

$$\frac{dr}{dt} = \frac{dh_1}{dt}$$

This time θ also tends to 0. Thus, giving us the principle as below:

$$\frac{v^2\cos^2\beta}{r} = r\left[\frac{d\theta}{dT}\right]^2$$

If we explore the last way of proving the above equation, we have:

$$\overline{OC}^2 = \overline{CD}^2 + \overline{OD}^2 - 2 * \overline{CD} * \overline{OD} * \cos\gamma$$

$$(R + h_1)^2 = v^2T^2 + (R + h_2 + h_3)^2 + 2v^2T^2(R + h_2 + h_3)\cos\gamma$$

Differentiating the above equation twice with respect to T and putting the limiting values we obtain again the equation as before and noting this time that:

$$\gamma \to (\pi - \alpha)$$

$$\frac{V^2\cos^2\beta}{r} = r\left[\frac{d\theta}{dT}\right]^2$$

The equation is lengthy so extra care should be exercised, and the following result should be used:

$$\frac{d\gamma}{dT} = -\frac{d\theta}{dT}$$

where $\gamma = \pi - (\theta + \alpha)$

In short, we have kept the following as variables and constants while using Alkashi Theorem

Constants: β, α, h_1, v
Variables: $T, h_2, h_3, \gamma, \theta$

Gravitation will not see such heights again after I am gone. Gravitivity is just 1/4th of gravitation as will become clear next evidently and abundantly.

Now if we reconsider Figure 3 with a little change but *heavier than mountains*, we have as below in figure 15:

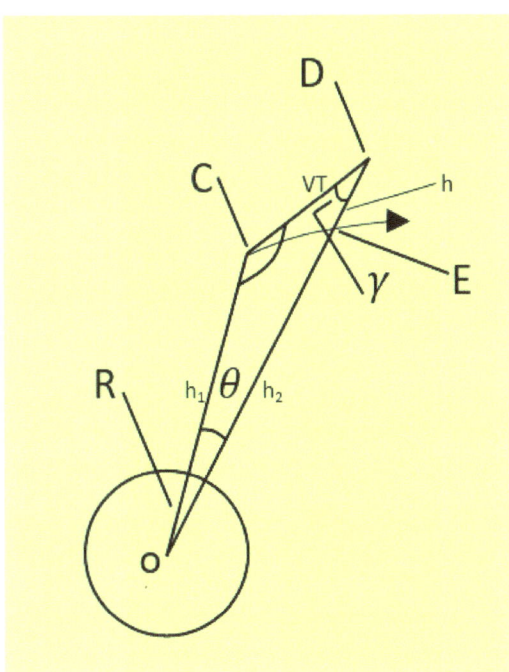

Figure 15. *Tahirian Inertial Centripetal/Centrifugal Force.*

The body moves under gravitation in a curved path shown above with an arrow. The tendency of inertia is to restore the body back to the matter-free straight path $\underset{CD}{\to}$ with *centrifugal acceleration* equal in magnitude to the gravitational acceleration but opposite in direction and vice versa when the main

switch of gravitation is turned off the body flies back to inertial motion in a straight line and is then opposed by a Centripetal force which balances the already present Centrifugal force. In inertial motion the Centripetal/Centrifugal forces exist as a couple in opposite directions. The only way it could do is by travelling from point E back to D with a velocity $V' = \frac{dh}{dT}$. So, keeping Figure 15 in mind, the equations will be changed a little with the addition of h as below,

$$\overline{OD}^2 = \overline{OC}^2 + \overline{CD}^2 - 2 * \overline{OC} * \overline{CD} * \cos\alpha$$

$$(\overline{OE} + \overline{DE})^2 = \overline{OC}^2 + \overline{CD}^2 - 2 * \overline{OC} * \overline{CD} * \cos\alpha$$

Distance h_2 is reduced by h accordingly.

$$(R + h_2 + h)^2 = (R + h_1)^2 + (VT)^2 - 2(R + h_1)VT\cos\alpha$$

Differentiating the above equation with respect to T we obtain

$$(R + h_2 + h)\left(\frac{dh_2}{dT} + \frac{dh}{dT}\right) = 0 + V^2T - (R + h_1)V\cos\alpha$$

$$(R + h_2 + h)\left(\frac{dh_2}{dT} + \frac{dh}{dT}\right) = V^2T - (R + h_1)V\cos\alpha$$

Now putting the limits i.e., when T tends to 0, h_2 tends to h_1 and also noting the fact that:

$$r' = R + h_1$$

$$\frac{dr'}{dT} = \frac{dh_1}{dT}$$

Now in the limiting case or instantaneous case, we obtain,

$$r\left(\frac{dr'}{dT} + V'\right) = 0 - rV\cos\alpha$$

$$\frac{dr'}{dT} + V' = +V\sin\beta$$

Broadly speaking, in both the two scenarios of the particle trajectory in inertial motion i.e., $\alpha = 90 \pm \beta$, the above equation is as follows:

$$\frac{dr'}{dT} + V' = \pm V\sin\beta$$

Where:

$$\frac{dh}{dT} = V'$$

If I differentiate the equation below once again, I obtain as follows:

$$\frac{d}{dT}\left[(R + h_2 + h)\left(\frac{dh_2}{dT} + \frac{dh}{dT}\right)\right] = \frac{d}{dT}[V^2T - (R + h_1)V\cos\alpha]$$

$$\left(\frac{dh_2}{dT} + \frac{dh}{dT}\right)^2 + (R + h_2 + h)\left(\frac{d^2h_2}{dT^2} + \frac{d^2h}{dT^2}\right) = V^2$$

Now putting the limits i.e., when T tends to 0, h_2 tends to h_1 and noting the fact that:

$$r = R + h_1$$

$$\frac{dr}{dT} = \frac{dh_1}{dT}$$

I obtain:

$$\left(\frac{dr'}{dT} + V'\right)^2 + r'\left(\frac{d^2r'}{dT^2} + \frac{dV'}{dT}\right) = V^2$$

Now we know from previous that:

$$\frac{dr'}{dT} + \mathbf{V'} = \pm V\sin\beta \text{ (both cases)}$$

So, putting the value we finally obtain as below:

$$\frac{d^2r'}{dT^2} + \frac{dV'}{dT} = \frac{V^2\cos^2\beta}{r'}$$

Space is a silhouette of gravitational communication between two pieces of matter. As soon as the matter pieces are put into it, it replaces the inertial centripetal acceleration as below with mechanical centripetal acceleration of gravity.

$$\frac{\mathbf{dV'}}{\mathbf{dT}} = \frac{\mathbf{GM}}{\mathbf{r^2}} = \frac{\mathbf{dV''}}{\mathbf{dT}}$$

When the inertial centripetal acceleration is replaced by the mechanical or physical acceleration due to communication at the speed of gravity, it is instantaneously (with speed of gravity) opposed by the other force of the couple, i.e., Centrifugal force. This is the tendency of inertia – the reactive centrifugal force. The associated velocities are as below:

$$V' = V\sin\beta - \frac{dr'}{dT}$$

$$V' = i^2\left[\frac{dr'}{dT} - V\sin\beta\right]$$

$$V' = i^2 V''$$

History will now drop the rotation as below:

$$V' = V''$$

Otherwise $\frac{\mathbf{dV'}}{\mathbf{dT}} = \mathbf{0}$

$$\frac{d^2r'}{dT^2} = \frac{V^2\cos^2\beta}{r'}$$
Where:

$$\frac{dr'}{dT} = \pm \pm iV\cos\beta$$

Instantaneous Gravitation $\equiv$ Inertia

$$\frac{dr'}{dT} \equiv \frac{dr}{dT}$$

Since gravitational communication happens at the speed of gravity which is incidentally the same value as that of the speed of light. **The rotation of components happens at the speed of gravity as well.** That's why the speed of gravity is inertial speed as well. Amazing, isn't it?

Humanity! lend me your ears. Gravity has cut $\overrightarrow{DE}$ out of the total portion of Inertia which is $\overrightarrow{OD}$. The chunk that the gravitational centripetal force has cut out of inertial Centripetal force *activates* an equal amount of counteractive force in the opposite direction as Centrifugal force. Basically

$$\frac{dV'}{dT} = \frac{dV''}{dT} = \frac{V^2\cos^2\beta}{r} - \frac{d^2r}{dT^2}$$

Centripetal and centrifugal forces exist as a couple in counter-action to each other. Gravity has cut an equal amount of centripetal force. Gravity can't cut more than what it has got the capacity to cut! Different capacities of gravities on the surfaces of different planets cut different chunks of inertial centripetal force and *activates* an equal amount of counteractive force in the opposite direction as Centrifugal force.

Inertia has got a reinforcement in case it is disturbed by gravities. That reinforcement is the redundant component in the non- radial direction as vcosβ. Since gravities are attractive forces and go through centre of masses, hence they are radial forces. They try to make changes in the radial direction. Nature has already provided or reinforced inertia with the provision of rotation of the redundant component by 90 degrees as ivcosβ. This component is ready to assist inertia to restore conditions back to inertia when perturbed by gravities. Since gravities are radial forces so this redundant component is rotated in the radial direction to assist inertia to produce centripetal/centrifugal forces of inertia. So, inertia has already been more than sufficiently provided with the provision or reinforcement or support to counteract gravity between masses. Amazing, isn't it?

Rotations in space happen in a similar way, though not congruent, as the rotation of the weight of fluid displaced in the upward direction as buoyant force. When the body was not immersed the force of the same amount of fluid, that would be displaced, was acting in the direction of gravity but as soon as the body is immersed in the fluid the fluid does a rotation of 180 degrees of the same amount of fluid displaced in the form of buoyant force which when subtracted from the weight of the object in air produces the apparent weight of the immersed object.

So far, we have talked about the tendency of inertia or the reactive power of inertia in response to active power of gravitation or simply the action of gravitation. There is a complex velocity and a real acceleration associated with inertia but equal to zero. For inertial motion, we have the condition:

$$V' = 0 \text{ (for inertial motion)}$$

Natural Theory of Relativity, Inertia, Gravitation & Gravitivity

$$\frac{dr}{dT} = \pm V\sin\beta$$

As a result of the above equation, we obtain as below and as went before:

$$\frac{d^2r}{dT^2} = \frac{V^2\cos^2\beta}{r}$$

The Tahirian Solution (TS) of the above differential equation remarkably comes out to be, for the first time in history, as

$$\frac{dr'}{dT} = \pm V\sin\beta$$

&

$$\frac{dr'}{dT} = \pm \pm iV\cos\beta$$

where $i = \sqrt{-1}$

$$h = rv\cos\beta$$

Now we see that this is another condition of inertia. It gives us another equation as below:

$$V' = V\sin\beta - \frac{dr'}{dT}$$

$$V' = \pm V\sin\beta - \pm \pm iV\cos\beta \ (= 0 \text{ for inertial motion})$$

$$V' = \pm V\sin\beta \mp \pm iV\cos\beta = \pm V\sin\beta \mp \pm iV\cos\beta$$

where $\frac{dr'}{dT} = \pm iV\cos\beta$

$r \to r' \to r$ Amazing, isn't it?

The combination of signs that we are interested in is the one that adds up the two components of velocity algebraically in magnitude either in the centrifugal direction or the centripetal direction. Other combinations are inadmissible.

$$V' = \pm V\sin\beta \pm \pm iV\cos\beta$$

Dropping the rotations along with their respective signs, we obtain as follows:

$$V' = \pm V\sin\beta \pm V\cos\beta$$

$$V' = +(V\sin\beta + V\cos\beta) \text{ (Tahirian Centripetal Velocity)}$$

$$V'' = -(V\sin\beta + V\cos\beta) \text{ (Tahirian Centrifugal Velocity)}$$

+i represents anticlockwise rotation of the component $v\cos\beta$ by 90 degrees and -i represents clockwise rotation of the component $v\cos\beta$ by 90 degrees in the direction of +$v\sin\beta$ and -$v\sin\beta$ respectively.

Let me differentiate the above expressions of the velocities to see what an amazing thing I get!

$$\frac{dV'}{dT} = +\frac{d(V\sin\beta + V\cos\beta)}{dT}$$

$$\frac{dV'}{dT} = \frac{d(V\sin\beta)}{dT} + \frac{d(V\cos\beta)}{dT}$$

Using the facts:

$$h = rv\cos\beta \ \& \ \frac{dr}{dT} = +V\sin\beta$$

I obtain,

$$\frac{dV'}{dT} = \frac{d^2r}{dT^2} - \frac{V^2\cos^2\beta}{r}$$

Which must be written as,

$$\frac{dV'}{dT} = i^2\left[\frac{V^2\cos^2\beta}{r} - \frac{d^2r}{dT^2}\right]$$

$$i^2\left[\frac{V^2\cos^2\beta}{r} - \frac{d^2r}{dT^2}\right] = \text{Centrifugal Force}$$

$$\frac{V^2\cos^2\beta}{r} - \frac{d^2r}{dT^2} = \text{Centripetal Force}$$

$i^2 = $ rotation of 180 degrees

Meaning simply that a couple of Centrifugal and Centripetal forces exist in nature! When one is rotated by 180 degrees, we get the other!

Action and reaction are equal in magnitude but opposite in direction.

Now rewrite the same equation a little bit differently as below:

$$\frac{dV'}{dT} = -\frac{d(-V\sin\beta - V\cos\beta)}{dT}$$

$$\frac{dV'}{dT} = -\frac{d(-V\sin\beta)}{dT} - \frac{d(-V\cos\beta)}{dT}$$

Using the facts:

$$\frac{dr}{dT} = -V\sin\beta; \ h = rv\cos\beta \ \& \frac{dr}{dT} = -V\cos\beta$$

$$V\sin\beta = \sqrt{v^2 - v^2\cos^2\beta}$$

I obtain,

$$\frac{dV'}{dT} = \frac{V^2\cos^2\beta}{r} - \frac{d^2r}{dT^2} \text{ (Centrifugal force)}$$

So, we see that:

$$\text{Centrifugal force} = \frac{V^2\cos^2\beta}{r} - \frac{d^2r}{dT^2} = \text{Centripetal force}$$

Amazing, isn't it?

Similarly, by using the centripetal velocity, we can prove the above results by differentiating in two different ways:

$$V'' = -(V\sin\beta + V\cos\beta)$$

$$V'' = -V\sin\beta - V\cos\beta$$

The rusk dipped and taken out and shaken over the evening cup of tea is an example of inertial motion. I wish humans knew it! Alas! They didn't know from the demise of Sir Ibn Bajjaj till the Advent of Dr.Tahir! Rusk and planets all constitute accelerated motion according to the fundamental Universal Principles of Tahir. Yeah! A car moving along a curved path with accelerated motion along the curve is another accelerated motion example. Yeah! car accelerating and decelerating in straight line motion is also an accelerated motion example again according to Tahirian Cosmology and much more.

In Tahirian Classical Celestial Mechanics the expression for the Tahirian Centrifugal Acceleration (TCA) and the corresponding force is pole apart different than the Notonian (Noton and accessories) enquiry into the Field of Islamic Celestial Mechanics by Ibne-Sina, Abul Barakat Al-Baghdadi, Ibne- Bajjaj and Janab Fayaz Tahir.

The centrifugal force acts in the direction of deceleration which is opposite to the direction of acceleration. Deceleration in one direction is accompanied by acceleration in the opposite direction. In other words, deceleration in one direction is equivalent to acceleration in the opposite direction. Deceleration and acceleration have opposite signs, but their magnitudes are the same. Take the example of a ball thrown in the air with speed less than escape. When it is going away from earth it is decelerating but at the same time it is accelerating towards the Centre of the earth. Why because after it takes the turn it returns towards the Centre of the earth with the same acceleration. So, both the motions of deceleration and acceleration are equivalent except the opposite signs. Ther is a very sublime and subtle blend of algebraic signs and Tahirian Space Rotations (TSR) and to see it is as difficult as seeing a black ant walking on a black wall on a black night!

When inertia is action, gravitation is reaction (ball thrown in the air with inertial velocity v). And when gravitation (centripetal acceleration) is an action then inertia is reaction -tendency of inertia to react. What inertia (reaction) does to gravitation (action), gravitation (reaction) does the same thing to inertia (action- a body thrown up from the surface of the earth with inertial velocity). The acceleration and the deceleration are equal in magnitude but opposite in directions. In other words, gravitation pulls a body with an acceleration (centripetal) which is the same as the deceleration with which inertia tends to decelerate the body in the same direction which in turn is the same as the same acceleration but in the opposite direction. Deceleration in one direction is the same as the same amount of acceleration but in the opposite direction.

When we apply brakes while driving a car in a straight line we decelerate. It is the inertial force, i.e., Tahir Centrifugal Force

$$m\frac{d^2r}{dT^2} \quad \text{where} \quad \frac{d^2r}{dT^2} < 0$$

that pushes us forward in the direction of deceleration. Centrifugal force acts on particles that have got some degree of freedom as compared to the particles that are rigidly attached to the vehicle undergoing centripetal acceleration. Centrifugal force is a reaction force to mechanical Centripetal force. That is why we have seat belts in our cars to keep us rigidly attached, to some extent, to the rest of the car and hence prevents us hitting the dashboard while going in the direction of deceleration. Centrifugal force vanishes as soon as the mechanical Centripetal force vanishes so that we obtain inertial tangential motion. However, when something is thrown up in the air, we apply mechanical Centrifugal force to it as action force. Consequently, the gravitational force acts in reaction to it as mechanical centripetal force.

In the City of Love, put thy name on the list-of-lovers.
Produce a new Era of new Mornings and new Evenings!

When we read World History, we see that these thoughts were first thought and written by the Muslim Fathers of Mechanics of Inertia and Gravitation namely Ibn-ur-Rushd Ibne-Sina, Abul-Barakat Al-Baghdadi, Al-Khazini, Ibne-Bajjaj etc. [2].

Now, let me explain the above with a slightly different approach using the equations below:

$$\frac{dr}{dT} = \pm V\sin\beta$$

$$\frac{d^2r}{dT^2} = \frac{V^2\cos^2\beta}{r}$$

$$\frac{dr}{dT} = \pm \pm iV\cos\beta$$

The first set of plus and minus signs is the algebraic one and the other set of plus and minus signs is the rotational or geometric one associated with the rotation of 90 degrees in the cosine term above. In other words, the second set of signs, geometric ones, shows either clockwise or anticlockwise rotation of the quantity associate with $i = \sqrt{-1}$.

$$h = +rV\cos\beta$$

In this approach , I won't consider instantaneous gravitation is equal to Inertia but rather just talk about inertia only.

We know that if c = a & c = b, then c = a + b, so we obtain as below:

$$\frac{dr}{dT} = \pm V\sin\beta \pm \pm iV\cos\beta$$

Dropping rotation of 90 degrees ($\equiv \sqrt{-1}$) in two opposite directions clockwise and anti-clockwise along with their two signs $\pm$, I obtain as below:

$$\frac{dr}{dT} = \pm V\sin\beta \pm V\cos\beta$$

$$\frac{dr}{dT} = \pm(V\sin\beta + V\cos\beta)$$

All other combinations of signs are inadmissible.

$$V' = +V\sin\beta + V\cos\beta \quad \text{(Centrifugal Velocity)}$$
$$V'' = -(V\sin\beta + V\cos\beta) \quad \text{(Centripetal Velocity)}$$

$$V' = i^2 V''$$

Now let me prove the Work-Energy Theorem in an unprecedented manner as below:

$$\frac{dv}{dT} = -\frac{GM}{r^2}\sin\beta$$

$$\frac{dv}{dr}\frac{dr}{dT} = -\frac{GM}{r^2}\sin\beta$$

$$\frac{dv}{dr}v\sin\beta = -\frac{GM}{r^2}\sin\beta$$

$$vdv = -\frac{GM}{r^2}dr$$

Integrating we have

$$\int_{v_1}^{v_2} vdv = \int_{r_1}^{r_2} -\frac{GM}{r^2}dr$$

$$\frac{v_2^2}{2} - \frac{v_1^2}{2} = GM\left(\frac{1}{r_2} - \frac{1}{r_1}\right)$$

Another remarkable way to prove the same as above is as below:

$$\frac{dv}{dT} = -\frac{GM}{r^2}\sin\beta$$

$$\frac{dv}{ds}\frac{ds}{dT} = -\frac{GM}{r^2}\sin\beta$$

$$\frac{dv}{ds}v = -\frac{GM}{r^2}\sin\beta$$

$$vdv = -\frac{GM}{r^2}\sin\beta ds$$

$$vdv = -\frac{GM}{r^2}dr$$

$$\int_{v_1}^{v_2} vdv = \int_{r_1}^{r_2} -\frac{GM}{r^2}dr$$

Now let us prove it in an Exotic Style never seen by eyes before the Advent *of Tahir* and more *intuitively* by equating the work done by gravitational force with the Tahir Centrifugal Force as below for the non-gravitivistic case:

$$\int \vec{F}.d\vec{r} = \text{workdone}$$

Vectors are a partial reality. We shouldn't be too much dependent upon vectors and should use them only if they give algebraically correct notions about Islamic Physics (Ultimate Truth).

$$\|d\vec{r}\| = dS \text{ (along the trajectory)} \neq dr$$

$$\int_{r_1}^{r_2} -\frac{Gm_1m_2}{r^2}dS\cos\left(\frac{\pi}{2}-\beta\right) = \int_{r_1}^{r_2} m_2\frac{d\vec{V'}}{dT}.d\vec{r}$$

$$\theta = \frac{\pi}{2} - \beta \text{ (particle going away from gravitating mass)}$$

$$\int_{r_1}^{r_2} -\frac{Gm_1m_2}{r^2}dS\sin\beta = \int_{r_1}^{r_2} m_2\frac{dV'}{dT}dS\cos\left(\frac{\pi}{2}-\beta\right)$$

$$= \int_{r_1}^{r_2} m_2\left(\frac{d^2r}{dT^2} - \frac{V^2\cos^2\beta}{r}\right)dS\sin\beta$$

$$-\int_{r_1}^{r_2} \frac{Gm_1m_2}{r^2}dr = m_2\int_{r_1}^{r_2}\left(\frac{d^2r}{dT^2} - \frac{V^2\cos^2\beta}{r}\right)dr$$

$$dr = dS\sin\beta$$

The above could also be very easily achieved by integrating both sides of the Tahirian Differential Equation of gravitation w.r.t. r as below:

$$-\frac{Gm_1m_2}{r^2} = \frac{d^2r}{dT^2} - \frac{V^2\cos^2\beta}{r}$$

But then I would not have been truthful to the height of truthfulness in calling my theory the "Mother of all Theories" (MOAT). I can't give a present to the naives (poking, anstan, noton, magllell etc.) of my Time and before better than Psychological MOAT.

The body is going away from the gravitating mass. The Tahirian Centrifugal Force (TCF) is in the direction of increasing r. So, a certain amount of work by an inertial force in one direction should have the same sign as the same amount of work in the same direction by a mechanical force.

$$-\int_{r_1}^{r_2} \frac{Gm_1m_2}{r^2} dr = m_2 \int_{r_1}^{r_2} \left(\frac{d}{dT}\frac{dr}{dT} - \frac{h^2}{r^3}\right) dr$$

$$= m_2 \int_{r_1}^{r_2} \frac{d}{dT}\frac{dr}{dT} dr - m_2 \int_{r_1}^{r_2} \frac{h^2}{r^3} dr$$

$$= m_2 \int_{r_1}^{r_2} d\left(\frac{dr}{dT}\right)\frac{dr}{dT} - m_2 \int_{r_1}^{r_2} \frac{h^2}{r^3} dr$$

$$= m_2 \int_{r_1}^{r_2} \frac{dr}{dT} d\left(\frac{dr}{dT}\right) - m_2 \int_{r_1}^{r_2} \frac{h^2}{r^3} dr$$

$$= m_2 \int_{v_1}^{v_2} v\sin\beta\, d(v\sin\beta) - m_2 \int_{r_1}^{r_2} \frac{h^2}{r^3} dr$$

$$= m_2 \left|\frac{v^2\sin^2\beta}{2}\right|_{v_1}^{v_2} - m_2 h^2 \int_{r_1}^{r_2} \frac{1}{r^3} dr$$

$$= m_2 \left|\frac{v^2\sin^2\beta}{2}\right|_{v_1}^{v_2} + m_2 \left|\frac{h^2}{2r^2}\right|_{r_1}^{r_2}$$

$$= m_2 \left|\frac{v^2\sin^2\beta}{2}\right|_{v_1}^{v_2} + m_2 \left|\frac{v^2\cos^2\beta}{2}\right|_{v_1}^{v_2}$$

$$= m_2 \left|\frac{v^2\sin^2\beta}{2} + \frac{v^2\cos^2\beta}{2}\right|_{v_1}^{v_2}$$

$$= m_2 \left|\frac{v^2}{2}\right|_{v_1}^{v_2} = m_2 \frac{v_2^2}{2} - m_2 \frac{v_1^2}{2} = Gm_1m_2\left(\frac{1}{r_2} - \frac{1}{r_1}\right)$$

$$m_2 \frac{v_2^2}{2} - m_2 \frac{v_1^2}{2} = Gm_1m_2\left(\frac{1}{r_2} - \frac{1}{r_1}\right)$$

Can I now say that Tahir's Anatomy of Gravitation and Inertia is your MOAT, my dear reader!

Has anyone ever thought and eventually produced a proof better than the above?

Now let me prove the same result in another exotic way and even more intuitively as below using Tahirian Centrifugal Velocity (TCV) $\mathbf{V'}$. Tahir Centrifugal Velocity has been eventually found in the Islamic (World) History for the first and the last time as below:

$$\int \vec{F}.d\vec{r} = \text{workdone}$$

$$\int_{r_1}^{r_2} m_2 \frac{dV'}{dT}.dr$$

$$\int_{r_1}^{r_2} m_2 \frac{dV'}{dT} dS\cos\theta \ \text{(as before)}$$

$$\int_{v_1''}^{v_2''} m_2 dV' \frac{dr}{dT}$$

$$V' = V\sin\beta - \frac{dr}{dT} = V\sin\beta - \pm iV\cos\beta \ \text{(TCV)}$$

$$V\sin\beta = \frac{dr}{dT} = \pm iV\cos\beta$$

The solution of the above equation is as below:

$$V = 0 \quad \textit{Another condition of Ibne-Sinian – Tahirian inertia}$$

$$\int_{v_1'}^{v_2'} m_2 \frac{dr}{dT} dV' = \int_{v_1'}^{v_2'} m_2(-V' + V\sin\beta) dV'$$

$$= \int_{v_1'}^{v_2'} m_2(-V' + V\sin\beta) dV' = m_2\left|-\frac{V'^2}{2} + V'V\sin\beta\right|_{v_1'}^{v_2'}$$

$$= m_2\left|-\frac{V'^2}{2} + V'V\sin\beta\right|_{V_1\sin\beta_1 - \pm iV_1\cos\beta_1}^{V_2\sin\beta_2 - \pm iV_2\cos\beta_2}$$

$$= m_2 \frac{v_2^2}{2} - m_2 \frac{v_1^2}{2}$$

The above is the expression for the difference in the Inertial Kinetic Energy of a body. The above derivation is very volatile and can sometimes disappear even during the process of derivation.

Speaking on Tahir Scale Consciousness (TSC) the above equation is basically part of a pair as will come next in Binary Star Systems and as follows in figure 16.

$$\frac{m_1 V_{12}^2}{2} - \frac{m_1 V_{11}^2}{2} = \frac{Gm_1m_2}{r_{12}t_1^2} - \frac{Gm_1m_2}{r_{11}t_1^2}$$

$$\frac{m_2 V_{22}^2}{2} - \frac{m_2 V_{21}^2}{2} = \frac{Gm_1m_2}{r_{22}t_2^2} - \frac{Gm_1m_2}{r_{21}t_2^2}$$

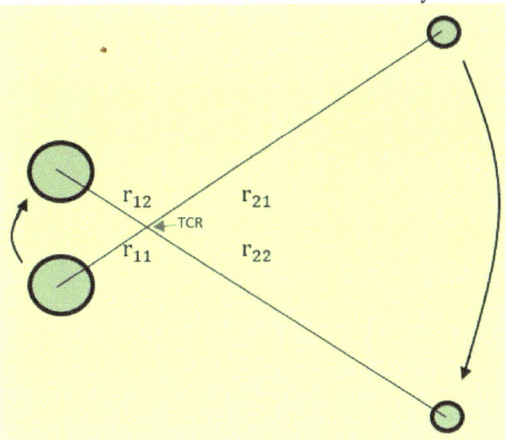

Figure 16. Tahirian Binaries- Stars, Planets and Projectiles.

The above pair is the non-gravitivistic binaries. The gravitivistic pair exists likewise and can be deducted from my article on binary star systems.

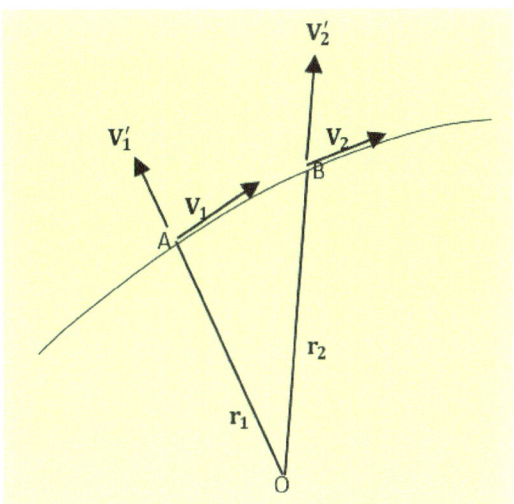

Figure 17. Tahirian Inertial Velocities in action at a point under Gravitation ($V_1 \neq V_2$).

$$\int \mathbf{F}.\,\mathbf{dr}$$

The above formula has been utilized by Jamanoo Cala Massiwalla in his commentary on wave theory. Meaning that this formula should be used where the intention is to either derive the speed of gravity as a result or derive the results for a physical quantity *without* the inclusion of the speed of gravity. Massiwala didn't know what he was doing and accidentally concluded the speed of the wave to be the speed of light instead of the speed of gravity. Speed of light is not an Inertial quantity because it undergoes deflection and is always under the sway of gravitivity one way or the other through the gravitational wave or graviton if it exists. Speed of light is a physical quantity only when it is considered in the presence of the speed of gravity. So, these integrals should not have concluded a physical quantity which itself deflects due to the speed of gravity. These line integrals of work done are inertial integrals that

calculate the values screened by the speed of gravity. Tahirian Gravitational Integral will come next where it utilizes the speed of gravity in it to find the kinetic energy of a body under gravitation because gravitation without Gravitivity is just screened acceleration, as in inertia, and nothing else. The vector fields and the Sibreen/geen and Shakeel/stok enquiry into the Islamic Mathematics deals with vector fields that are not gravitational and don't have any gravitational speed associated with them at all.

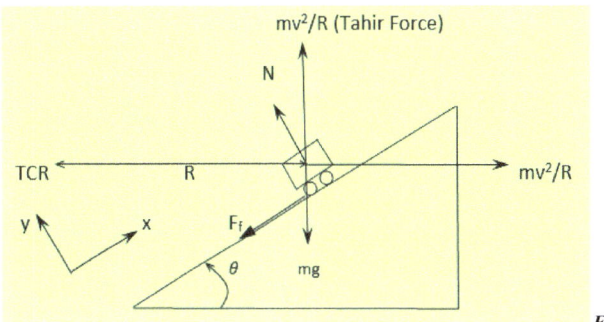

Figure 18. Banked Turn. Tahir Superelevation Design of Highways.

Now let Fayaz Tahir do the derivation of the maximum speed of vehicles on superelevated banked turns on highways/motorways.

Summing up forces in the y direction I have

$$N + m\frac{v_{max}^2}{R}\cos\theta - mg\cos\theta - m\frac{v_{max}^2}{R}\sin\theta = 0$$

Summing up forces in the x direction I have

$$m\frac{v_{max}^2}{R}\cos\theta + m\frac{v_{max}^2}{R}\sin\theta - mg\sin\theta - F_f = 0$$

$$N = m\frac{v_{max}^2}{R}[\sin\theta - \cos\theta] + mg\cos\theta$$

$$F_f = \mu N$$

$$F_f = \mu N = \mu m\frac{v_{max}^2}{R}[\sin\theta - \cos\theta] + \mu mg\cos\theta$$

Substituting the value of the friction force into the second equation above, I have

$$m\frac{v_{max}^2}{R}\cos\theta + m\frac{v_{max}^2}{R}\sin\theta - mg\sin\theta - F_f = 0$$

$$m\frac{v_{max}^2}{R}\cos\theta + m\frac{v_{max}^2}{R}\sin\theta - mg\sin\theta = F_f$$

$$m\frac{v_{max}^2}{R}\cos\theta + m\frac{v_{max}^2}{R}\sin\theta - mg\sin\theta$$
$$= \mu_s m\frac{v_{max}^2}{R}[\sin\theta - \cos\theta] + \mu_s mg\cos\theta$$

Separating the cosine and sine terms I have

$$\tan\theta = \frac{\mu_s g - \frac{\mu v_{max}^2}{R} - \frac{v_{max}^2}{R}}{\frac{v_{max}^2}{R} - g - \frac{v_{max}^2}{R}}$$

$$v_{max} = \sqrt{\frac{gR(\tan\theta + \mu_s)}{(1 - \mu_s)\tan\theta + (1 + \mu_s)}}$$

Similarly, I can find the minimum velocity required on the bank by considering the friction force in the other direction as follows:

$$v_{min} = \sqrt{\frac{gR(\tan\theta - \mu_s)}{(1 + \mu_s)\tan\theta + (1 - \mu_s)}}$$

For the Wall of Death (Well of Death) at $\theta = 90°$, *only I have* the maximum and minimum velocities *in the world* as below:

$$v_{max} = \sqrt{\frac{gR}{1 - \mu_s}} \quad \& \quad v_{min} = \sqrt{\frac{gR}{1 + \mu_s}}$$

If there is a traffic jam on the banked curve due to an incident ahead or some other reason, then, the cars velocities must come to zero. Therefore, the condition for the minimum velocity/speed should be as below:

$$v_{min} = \sqrt{\frac{gR(\tan\theta_{max} - \mu_s)}{(1 + \mu_s)\tan\theta_{max} + (1 - \mu_s)}} = 0$$

$$\tan\theta_{max} - \mu_s = 0$$

$$\tan\theta_{max} = \mu_s$$

$$\theta_{max} = \tan^{-1}\mu_s$$

So, θ must be less than $\tan^{-1}\mu_s$, in order, for cars NOT to slide upon each other towards the down end of the banked turn during a complete traffic jam, like in the wall of death ($\theta = 90° > \tan^{-1}\mu_s$) where we need a minimum speed to stay on the wall because the friction in the opposite direction of falling would NOT be sufficiently enough to hold the vehicle from sliding or falling. Similarly minimum velocities can't be maintained on the banked turn in the case of total traffic jam when:

$$\theta > \tan^{-1}\mu_s$$

Since minimum speed limits should NEVER ever be considered for design purposes, hence the maximum limit on θ - the angle of banked turn is:

$$\theta_{max} = \tan^{-1}\mu_s$$

The above values of minimum and maximum velocities are *also* the values for the condition of **skidding** of the vehicle either towards the upside of the banked turn or the downside.

However, after incorporating the minimum velocity condition for total traffic jam and replacing coefficient of static friction with the coefficient of kinetic friction, I obtain as below:

$$0 \le v \le \sqrt{\frac{gR(\tan\theta + \mu_k)}{(1 - \mu_k)\tan\theta + (1 + \mu_k)}}$$

$$\theta_{max} < \tan^{-1}\mu_k$$

In short, we want to utilize as much angle of superelevation as possible to let cars move at higher speeds around the banked turn without skidding, up or down end of the banked turn, and overturning but at the same time we don't want to give as much angle as it will cause the cars to skid towards the down end of the road in the case of total traffic jam!

To do the overturning analysis of the vehicles, particularly high trucks, I need to consider two important dimensions of the vehicle from the centre of gravity of the vehicle, i.e., b & d as in the figure below:

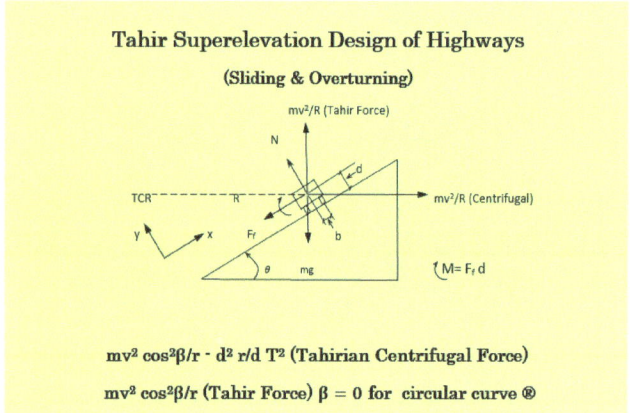

Figure 18a. *Banked Turn. Tahir Superelevation Design of Highways. Sliding and overturning of vehicle.*

For overturning, we will consider the condition as soon as the vehicle down tyre loses contact with the surface of the road. When this happens the normal reaction of the road is transmitted to the vehicle through the up tyre in contact with the surface of the road. Therefore, taking turning moments around the point of contact between the up tyre and the surface of the road, I obtain as below:

$$m\frac{v_{max}^2}{R}(b + d\tan\theta)\cos\theta + m\frac{v_{max}^2}{R}(d\sec\theta - (b + d\tan\theta)\sin\theta) = mg(b + d\tan\theta)\cos\theta$$

From Figure 18a, we see that only two moment arms are needed to take moments about the turning point, i.e., the right tyre of the vehicle as shown in the figure above and as below:

$$(b + d\tan\theta)\cos\theta \quad \& \quad d\sec\theta - (b + d\tan\theta)\sin\theta$$

Solving for v_{max},

$$v_{max} = \sqrt{gR\frac{(b\cos\theta + d\sin\theta)}{(d + b)\cos\theta + (d - b)\sin\theta}}$$

4. Binary star systems

To complete the story of Tahirian Binaries, we need to accomplish complete theory of gravitivity. Hence, let Tahir prove the relativistic time dilation formula with one clock in history once and forever. I will utilize figure 2 in the beginning that we used to prove the formula before. Since I am using only one clock, therefore, I will use t in place of t_1. Using the Theorem de Alkashi on triangle AEC, we have:

$$\overline{EC}^2 = \overline{AE}^2 + \overline{AC}^2 - 2 * \overline{AE} * \overline{AC} * \cos\psi$$

Angle EAC is ψ because the second clock at F is assumed to have converged to the first clock. Clock at F converges to clock at E to form one clock.

$$(c\delta t_1)^2 = a^2 + (vt + v\delta t)^2 - 2a(vt + v\delta t)\cos\psi$$

$$(c(T - t))^2 = a^2 + (vT)^2 - 2a(vT)\cos\psi$$

Differentiating the above equation w.r.t T we obtain as below:

$$2c^2(T - t) = 0 + 2v^2T - 2av\cos\psi$$

$$2c^2T - 2c^2t = 2v^2T - 2av\cos\psi$$

$$2c^2T - 2v^2T = 2c^2t - 2av\cos\psi$$

$$c^2T - v^2T = c^2t - av\cos\psi$$

$$c^2T(1 - \frac{v^2}{c^2}) = c^2t - av\cos\psi$$

$$T(1 - \frac{v^2}{c^2}) = t - \frac{av\cos\psi}{c^2}$$

$$T = \frac{t - \frac{av\cos\psi}{c^2}}{(1 - \frac{v^2}{c^2})}$$

Now, replacing a by r we obtain as follows:

$$T = \frac{t - \frac{rv\cos\psi}{c^2}}{(1 - \frac{v^2}{c^2})} = \frac{T'}{(1 - \frac{v^2}{c^2})}$$

$$T = \frac{T'}{(1 - \frac{v^2}{c^2})}$$

Differentiating the above equation w.r.t. T', we have

$$dT = \frac{dT'}{(1 - \frac{v^2}{c^2})}$$

Where:

$$T' = t - \frac{rv\cos\psi}{c^2}$$

Let $\psi = \frac{\pi}{2} + \beta$

$$T' = t + \frac{rv\sin\beta}{c^2}$$

I'll differentiate the above Tahirian Equation as below one by one by T', T & t

$$\frac{dT'}{d(T,T' \& t)} = \frac{d}{d(T,T' \& t)}(t + \frac{rv\sin\beta}{c^2})$$

$$dS = vdT = v'dT' = \forall dt \text{ (TIDLE)}$$

V is basically equal to

$$v_1 + v_2$$
$$\therefore$$
$$v = v_1 + v_2$$

It will be abundantly shown to be correct later in My Theory of Gravitation of Binary Stars.

Now, differentiating the above equation w.r.t. T, we have

$$\frac{dT'}{dT} = \frac{dt}{dT} + \frac{d}{dT}(\frac{rv\sin\beta}{c^2})$$

$$1 - \frac{v^2}{c^2} = \frac{dt}{dT} - \frac{v}{c^2}\frac{d}{dT}(r\sin\beta)$$

Now using the laws of the Tahirian Inertial Differential Line Element (TIDLE) dS as given above we have

$$\frac{dr}{dT} = v\sin\beta \qquad dr = vdT\sin\beta$$

$$dr = dS\sin\beta = (vdT)\sin\beta = (v'dT')\sin\beta = (\forall dt)\sin\beta$$

$$\frac{d\beta}{dT} = \frac{v\cos\beta}{r}$$

$$1 - \frac{v^2}{c^2} = \frac{dt}{dT} - \frac{v}{c^2}\frac{d}{dT}(r\sin\beta)$$

$$1 - \frac{v^2}{c^2} = \frac{dt}{dT} - \frac{v^2}{c^2}$$

$$1 - 2\frac{v^2}{c^2} = \frac{dt}{dT}$$

$$\frac{dT}{dt} = \frac{1}{(1 - 2\frac{v^2}{c^2})}$$

Now, differentiating the above equation w.r.t. T', we have

$$\frac{dT'}{dT'} = \frac{dt}{dT'} + \frac{d}{dT'}(\frac{rv\sin\beta}{c^2})$$

$$1 = \frac{dt}{dT'} - \frac{v}{c^2}\frac{d}{dT'}(r\sin\beta)$$

Now using the laws of the Tahirian Inertial Differential Line Element (TIDLE) dS as before we have

$$\frac{dr}{dT'} = v'\sin\beta \qquad dr = v'dT'\sin\beta \qquad dr = dS\sin\beta$$

$$\frac{d\beta}{dT'} = \frac{v'\cos\beta}{r} \qquad d\beta = \frac{dS\cos\beta}{r}$$

$$1 = \frac{dt}{dT'} - \frac{v}{c^2}\frac{d}{dT'}(r\sin\beta)$$

$$1 = \frac{dt}{dT'} - \frac{vv'}{c^2}$$

$$1 - \frac{vv'}{c^2} = \frac{dt}{dT'}$$

$$\frac{dT'}{dt} = \frac{1}{(1 - \frac{vv'}{c^2})}$$

Inserting the value of v' we obtain as below:

$$\frac{dT'}{dt} = \frac{1 - \frac{v^2}{c^2}}{(1 - 2\frac{v^2}{c^2})}$$

$$\frac{dT}{dt} = \frac{1}{(1 - 2\frac{v^2}{c^2})}$$

Now, differentiating the above equation w.r.t. t, we have

$$\frac{dT'}{dt} = \frac{dt}{dt} + \frac{d}{dt}(\frac{rv\sin\beta}{c^2})$$

$$\frac{dT'}{dt} = 1 + \frac{v}{c^2}[r\frac{d}{dt}\sin\beta + \sin\beta\frac{d}{dt}r]$$

$$\frac{dT'}{dt} = 1 + \frac{v}{c^2}[r\frac{d}{dt}\sin\beta + \sin\beta\frac{d}{dt}r]$$

$$\frac{dT'}{dt} = 1 + \frac{v}{c^2}\left[r\frac{d}{dt}\sin\beta + \sin\beta\frac{d}{dt}r\right]$$

$$\frac{dT'}{dt} = 1 + \frac{v}{c^2}[r\cos\beta\frac{d\beta}{dt} + \sin\beta\frac{dr}{dt}]$$

Differentiating the equation below w.r.t t we obtain

$$h = r\mathbf{v}\cos\beta$$

$$\frac{dh}{dt} = 0 = \frac{d}{dt}(r\mathbf{v}\cos\beta) = \mathbf{v}\frac{d}{dt}(r\cos\beta)$$

$$\frac{d}{dt}(r\cos\beta) = 0$$

$$r\frac{d}{dt}\cos\beta + \cos\beta\frac{dr}{dt} = 0$$

$$-r\sin\beta\frac{d\beta}{dt} + \cos\beta\frac{dr}{dt} = 0$$

$$\frac{d\beta}{dt} = \frac{\mathbf{v}\cos\beta}{r} \quad \& \quad \frac{dr}{dt} = \mathbf{v}\sin\beta$$

Inserting the above values, we have

$$\frac{dT'}{dt} = 1 + \frac{v}{c^2}[\mathbf{v}\cos^2\beta + \mathbf{v}\sin^2\beta]$$

$$\frac{dT'}{dt} = 1 + \frac{v\mathbf{v}}{c^2}$$

$$\mathbf{v} = \frac{v}{1 - 2\frac{v^2}{c^2}}$$

$$\frac{dT'}{dt} = \frac{1 - \frac{v^2}{c^2}}{(1 - 2\frac{v^2}{c^2})}$$

$$\frac{dT}{dt} = \frac{1}{(1 - 2\frac{v^2}{c^2})}$$

Putting the above value back in the previous equation we obtain

$$T = \frac{(t - \frac{rv\cos\psi}{c^2})}{\left(1 - \frac{v^2}{c^2}\right)} = \frac{T'}{\left(1 - \frac{v^2}{c^2}\right)}$$

$$dT = \frac{dT'}{\left(1 - \frac{v^2}{c^2}\right)} = \frac{dt}{\left(1 - 2\frac{v^2}{c^2}\right)}$$

The above are the fundamental equations of Tahirian Time dilations.

From the above equations of two different times, we will derive the precession of the perihelion of planets as follows.

$$dS = VdT = V'dT' = ɏdt$$

$$v' = \frac{v}{1 - \frac{v^2}{c^2}}$$

$$ɏ = \frac{V}{1 - 2\frac{v^2}{c^2}}$$

Differentiating the above first Equation, we obtain as below:

$$dv' = \frac{1 + \frac{v^2}{c^2}}{\left[1 - \frac{v^2}{c^2}\right]^2} dv$$

Or

$$\frac{dv'}{dT'} = \frac{1 + \frac{v^2}{c^2}}{\left[1 - \frac{v^2}{c^2}\right]^2} \frac{dv}{dT'}$$

Now, the above bridging equation can also be written as the set of the following equations.

$$\frac{d^2r'}{dT'^2} - \frac{v'^2\cos^2\beta'}{r'} = -\frac{GM}{r'^2}\left\{\frac{1+\frac{v^2}{c^2}}{\left[1-\frac{v^2}{c^2}\right]^2}\right\}; \quad h' = r'v'\cos\beta'$$

&

$$\frac{dr'}{dT'} = v'\sin\beta'$$

The above set of equations produces the result as below and as before.

$$\frac{dv'}{dT'} = \frac{1 + \frac{v^2}{c^2}}{\left[1 - \frac{v^2}{c^2}\right]^2} \frac{GM}{r'^2}\sin\beta'$$

An extremely important reality to note here at this juncture of History is that the term:

$$\frac{dV}{dT'} = \frac{GM}{r'^2}\sin\beta'$$

is NOT at all associated with static orbits!!! But rather it is the term, dimensionally similar, to the term of the static orbits, if there exists one! It is basically a dimensionally similar looking term BUT for the *partially* precessing orbits undergoing precession in the forward direction of motion due to proper Time Dilation *only*.

It needs to be clearly understood that the precession of the planets is, an absolutely, inertial effect related to the two distinct times with two distinct constant velocities associated with them. So, doing the barbecue of the first term with the help of chain rule we obtain,

$$\frac{d^2r'}{dT^2} = \frac{d}{dT}\left\{\frac{dr'}{dT'}\frac{dT'}{dT}\right\} = \frac{d}{dT'}\left\{\frac{dr'dT'}{dT'dT}\right\}\frac{dT'}{dT}$$

$$\frac{d^2r'}{dT^2} = \left(1 - \frac{v^2}{c^2}\right)\frac{d}{dT'}\left\{\frac{dr'dT'}{dT'dT}\right\} = \left(1 - \frac{v^2}{c^2}\right)^2\frac{d^2r'}{dT'^2}$$

$$\frac{d^2r'}{dT^2} = \left(1 - \frac{v^2}{c^2}\right)^2\frac{d^2r'}{dT'^2}$$

Where:

$$\frac{d}{dT'}\left\{\frac{dT'}{dT}\right\} = \frac{d}{dT'}\left(1 - \frac{v^2}{c^2}\right) = 0 \quad \textbf{(Gravitivity/relativity)}$$

We can also write the above equation as

$$\frac{1}{\left(1-\frac{v^2}{c^2}\right)^2}\frac{d^2r'}{dT^2}=\frac{d^2r'}{dT'^2}$$

Inserting the values in the equation below, we obtain

$$\frac{d^2r'}{dT'^2}-\frac{v'^2\cos^2\beta'}{r'}=-\frac{GM}{r'^2}\left\{\frac{1+\frac{v^2}{c^2}}{\left[1-\frac{v^2}{c^2}\right]^2}\right\}$$

$$\frac{1}{\left(1-\frac{v^2}{c^2}\right)^2}\frac{d^2r'}{dT^2}-\frac{v^2\cos^2\beta'}{\left(1-\frac{v^2}{c^2}\right)^2 r'}=-\frac{GM}{r'^2}\left\{\frac{1+\frac{v^2}{c^2}}{\left[1-\frac{v^2}{c^2}\right]^2}\right\}$$

Cancelling equal terms on both sides we obtain as below:

$$\frac{d^2r'}{dT^2}-\frac{v^2\cos^2\beta'}{r'}=-\frac{GM}{r'^2}\left\{1+\frac{v^2}{c^2}\right\}$$

At the end of the day, we came back from where we started with i.e.,

$$dS = VdT$$

No if we differentiate the equation as below:

$$\bar{V}=\frac{v}{1-2\frac{v^2}{c^2}}$$

We obtain:

$$d\bar{V}=\frac{1+2\frac{v^2}{c^2}}{\left[1-2\frac{v^2}{c^2}\right]^2}dv$$

Or

$$\frac{d\bar{V}}{dt}=\frac{1+2\frac{v^2}{c^2}}{\left[1-2\frac{v^2}{c^2}\right]^2}\frac{dv}{dt}$$

Now, the above bridging equation can also be written as the set of the following equations.

$$\frac{d^2\bar{r}}{dt^2}-\frac{\bar{V}^2\cos^2\beta}{\bar{r}}=-\frac{GM}{\bar{r}^2}\left\{\frac{1+2\frac{v^2}{c^2}}{\left[1-2\frac{v^2}{c^2}\right]^2}\right\};\ \hbar=\bar{r}\bar{V}\cos\beta$$

&

$$\frac{d\bar{r}}{dt}=\bar{V}\sin\beta$$

The above set of equations produces the result as below:

$$\frac{d\bar{V}}{dt}=\frac{1+2\frac{v^2}{c^2}}{\left[1-2\frac{v^2}{c^2}\right]^2}\frac{GM}{\bar{r}^2}\sin\beta$$

An extremely important reality to note here at this juncture of History is that the term:

$$\frac{dv}{dt}=\frac{GM}{\bar{r}^2}\sin\beta$$

is NOT at all associated with static orbits!!! But rather it is the term, dimensionally similar, to the term of the static orbits if there exists one! It is basically the dimensionally similar looking term BUT for the *partially* precessing orbits undergoing precession in the forward direction of motion due to the time it takes to bring information at the speed of gravity from one frame of reference to another *only*.

It needs to be clearly understood that the precession of the planets is an, absolutely, inertial effect related to the two distinct times with constant velocity in them. So, doing the barbecue of the first term with the help of chain rule we obtain,

$$\frac{d^2\bar{r}}{dT^2}=\frac{d}{dT}\left\{\frac{d\bar{r}}{dt}\frac{dt}{dT}\right\}=\frac{d}{dt}\left\{\frac{d\bar{r}dt}{dtdT}\right\}\frac{dt}{dT}$$

$$\frac{d^2\bar{r}}{dT^2}=\left(1-2\frac{v^2}{c^2}\right)\frac{d}{dt}\left\{\frac{d\bar{r}dt}{dtdT}\right\}=\left(1-2\frac{v^2}{c^2}\right)^2\frac{d^2\bar{r}}{dt^2}$$

$$\frac{d^2\bar{r}}{dT^2}=\left(1-2\frac{v^2}{c^2}\right)^2\frac{d^2\bar{r}}{dt^2}$$

Where:

$$\frac{d}{dt}\left\{\frac{dt}{dT}\right\}=\frac{d}{dt}\left(1-2\frac{v^2}{c^2}\right)=0\quad\textbf{(Gravitivity/relativity)}$$

We can also write the above equation as

$$\frac{1}{\left(1-2\frac{v^2}{c^2}\right)^2}\frac{d^2\bar{r}}{dT^2}=\frac{d^2\bar{r}}{dt^2}$$

$$\frac{d^2\bar{r}}{dt^2}-\frac{\bar{V}^2\cos^2\beta}{\bar{r}}=-\frac{GM}{\bar{r}^2}\left\{\frac{1+2\frac{v^2}{c^2}}{\left[1-2\frac{v^2}{c^2}\right]^2}\right\}$$

Inserting the values in the equation below, we obtain

Natural Theory of Relativity, Inertia, Gravitation & Gravitivity

$$\frac{d^2\digamma}{dt^2} - \frac{\Psi^2\cos^2\beta}{\digamma} = -\frac{GM}{\digamma^2}\left\{\frac{1 + 2\frac{v^2}{c^2}}{\left[1 - 2\frac{v^2}{c^2}\right]^2}\right\}$$

$$\frac{1}{\left(1 - 2\frac{v^2}{c^2}\right)^2}\frac{d^2\digamma}{dT^2} - \frac{\Psi^2\cos^2\beta}{\left(1 - 2\frac{v^2}{c^2}\right)^2\digamma} = -\frac{GM}{\digamma^2}\left\{\frac{1 + 2\frac{v^2}{c^2}}{\left[1 - 2\frac{v^2}{c^2}\right]^2}\right\}$$

Cancelling equal terms on both sides we obtain as below:

$$\frac{d^2\digamma}{dT^2} - \frac{v^2\cos^2\beta}{\digamma} = -\frac{GM}{\digamma^2}\left\{1 + 2\frac{v^2}{c^2}\right\}$$

At the end of the day, we came back from where we started with i.e., the lion came back to its den:

$$dS = VdT$$

Combining the two equations together below as

$$\frac{d^2r'}{dT^2} - \frac{v^2\cos^2\beta'}{r'} = -\frac{GM}{r'^2}\left\{1 + \frac{v^2}{c^2}\right\}$$

$$\frac{d^2\digamma}{dT^2} - \frac{v^2\cos^2\beta}{\digamma} = -\frac{GM}{\digamma^2}\left\{1 + 2\frac{v^2}{c^2}\right\}$$

Now, it is to be understood that there is only one orbit. So, in the limiting case of adding the gravitivistic terms together, we finally obtain the "Close Form" of the Gravitivistic force of Tahirian Gravitation that eyes were dying to see for the last one hundred years as below:

$$\frac{d^2r}{dT^2} - \frac{v^2\cos^2\beta}{r} = -\frac{GM}{r^2}\left\{1 + 3\frac{v^2}{s^2}\right\}$$

Both radial distances merge to r in the limiting case!

If c = a & c is also equal to b then c = a + b

The equation of the static orbit remains the same, what is added is the combined effect of proper time dilation and coordinate time delay to produce the overall Precession of the planets. Tahirian Precession of Planets (TPP) is the additive effect of the precessions due to Proper time Dilation and Coordinate Time Delay when receiving the information with the speed of gravitational information s. *One thing to clarify is that Tahir has projected the velocities v' & ¥ onto the Tahir Inertial Differential Line Element (TIDLE) dS so that their derivatives w.r.t. times can be easily equated to two different gravitational terms. Hence, they are the projected velocities!*

Solving the set of two Tahirian Equations as below by Tahir's Rule as went before, we have:

$$\frac{d^2r}{dT^2} - \frac{v^2\cos^2\beta}{r} = -\frac{GM}{r^2}\left\{1 + 3\frac{v^2}{s^2}\right\}$$

$$h = rv\cos\beta \quad \& \quad \frac{dr}{dT} = V\sin\beta$$

We obtain:

$$\frac{dv}{dT} = -\left\{1 + 3\frac{v^2}{s^2}\right\}\frac{GM}{r^2}\sin\beta$$

This can be easily integrated to produce what eyes of the people were dying to see for the last three hundred years as below. Applying Chain Rule, we obtain the integral

$$\int_{v_R}^{v_r}\frac{vdv}{\left\{1 + 3\frac{v^2}{s^2}\right\}} = -\int_{R}^{r}\frac{GM}{r^2}dr$$

$$\frac{s^2}{6}\int_{v_R}^{v_r}\frac{d(1 + 3\frac{v^2}{s^2})}{\left\{1 + 3\frac{v^2}{s^2}\right\}} = -\int_{R}^{r}\frac{GM}{r^2}dr$$

$$\int_{v_R}^{v_r}\frac{d(1 + 3\frac{v^2}{s^2})}{\left\{1 + 3\frac{v^2}{s^2}\right\}} = -\frac{6}{s^2}\int_{R}^{r}\frac{GM}{r^2}dr$$

$$\ln\left(\frac{1 + 3\frac{v_r^2}{s^2}}{1 + 3\frac{v_R^2}{s^2}}\right) = \frac{6GM}{s^2}\left(\frac{1}{r} - \frac{1}{R}\right)$$

$$\frac{1 + 3\frac{v_r^2}{s^2}}{1 + 3\frac{v_R^2}{s^2}} = e^{\frac{6GM}{s^2}\left(\frac{1}{r} - \frac{1}{R}\right)}$$

$$1 + 3\frac{v_r^2}{s^2} = \left(1 + 3\frac{v_R^2}{s^2}\right)e^{\frac{6GM}{s^2}\left(\frac{1}{r} - \frac{1}{R}\right)}$$

$$v_r^2 = \frac{s^2}{3}\left\{\left(1 + 3\frac{v_R^2}{s^2}\right)e^{\frac{6GM}{s^2}\left(\frac{1}{r} - \frac{1}{R}\right)} - 1\right\}$$

We will see next that another Tahir Constant of Binary Systems t can be added in the above equation as below:

$$v_{12}^2 = \frac{s^2}{3t_1^2}\left\{\left(1 + 3t_1^2\frac{v_{11}^2}{s^2}\right)e^{\left[\frac{6Gm_2}{s^2}\left(\frac{1}{r_{12}} - \frac{1}{r_{11}}\right)\right]} - 1\right\}$$

Setting $v_{12} = 0$ & $r_{12} = \infty$, we obtain the Tahirian Escape Speed (TES) as below:

$$v_{11} = \frac{s}{t_1}\sqrt{\frac{1}{3}\left(e^{\frac{6Gm_2}{s^2r_{11}}} - 1\right)}$$

Where $t_1 = 1 + \frac{m_1}{m_2}$

We can find the escape speed of the moon escaping the gravitational pull of the earth to be as above, where m_1 is the mass of moon and m_2 is the mass of earth. Escape speed is dependent upon both the masses involved! Haven't I seen a bigger picture of gravitation? Where are your second class, if not third, hubble and bubble telescopes?

We will utilize the above fundamental Tahirian Work-Energy formula in the coming Tahir Field Equations (TFEs) to find the precession of the planets including the Tahirian Binary Star-System Constant t.

Now let **Fayaz Tahir** find the binary equations of binary star systems.

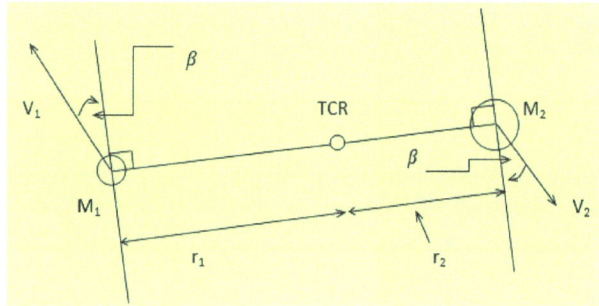

Figure 19. $\beta_1 = \beta_2 = \beta$ *in Binary Star Systems all over in the Tahirian Cosmos*

From the above figure let me assume that two particles of masses m_1 and m_2 are moving inertially with respect to each other making angles $\beta_1 = \beta$ and $\beta_2 = \beta$ as shown in the above figure. The relative velocity of the system is representative by v, the relative mass by m and the distance between their centers of mass as r. Hence, we have:

$$v^2 = (v_1 \cos\beta_1 + v_2 \cos\beta_2)^2 + (v_1 \sin\beta_1 + v_2 \sin\beta_2)^2$$

Simplifying further we obtain as below:

$$v^2 = v_1^2 + v_2^2 + 2v_1 \cos\beta_1 v_2 \cos\beta_2 + 2v_1 \sin\beta_1 v_2 \sin\beta_2$$

For two masses m_1 and m_2 to keep on moving in opposite directions but in parallel lines (we will prove it soon after this) under Tahir's command in physics i.e., Tahir Hypothetical Inertial Screening (THIS) as compared to its one case of Tahir Hypothetical Inertial Catastrophe (THIC). In THIS both masses are screened from each other's gravitational pull whereas in its one case of THIC we assume the catastrophe happens and the Sun vaporizes. However, in the figure above the two masses are not screened and they are rotating with respect to Tahir Center of Rotation (TCR) as below

$$\frac{v_1 \cos\beta_1}{r_1} = \frac{v_2 \cos\beta_2}{r_2} = \omega = \frac{d\theta}{dT}$$

Where θ is the angle of rotation of the binary-star systems around TCR.

$$\frac{r_1 v_1 \cos\beta_1}{r_1^2} = \frac{r_2 v_2 \cos\beta_2}{r_2^2}$$

$$\frac{h_1}{r_1^2} = \frac{h_2}{r_2^2}$$

$$\frac{r_1}{r_2} = \frac{\sqrt{h_1}}{\sqrt{h_2}}$$

$$r_1 = \frac{\sqrt{h_1}}{\sqrt{h_2}} r_2$$

Differentiating the above equation with respect to T, we obtain as below:

$$\frac{dr_1}{dT} = \frac{\sqrt{h_1}}{\sqrt{h_2}} \frac{dr_2}{dT}$$

Or

$$v_1 \sin\beta_1 = \sqrt{\frac{h_1}{h_2}} v_2 \sin\beta_2$$

$$\frac{v_1}{v_2} = \frac{\sqrt{h_1}}{\sqrt{h_2}}$$

Now substituting the above values systematically into the equation below

$$v^2 = v_1^2 + v_2^2 + 2v_1 \cos\beta_1 v_2 \cos\beta_2 + 2v_1 \sin\beta_1 v_2 \sin\beta_2$$

We obtain two equations as below,

$$v^2 = v_1^2 + v_2^2(1 + 2\sqrt{\frac{h_1}{h_2}})$$

And

$$v^2 = v_1^2(1 + 2\sqrt{\frac{h_2}{h_1}}) + v_2^2$$

Comparing the two equations for v^2 and simplifying, we obtain as below:

$$\frac{v_1}{v_2} = \frac{\sqrt{h_1}}{\sqrt{h_2}}$$

Which implies the result that $\beta_1 = \beta_2$. Now the same result could be proven very easily by dividing the two equations as below:

$$v_1 \sin\beta_1 = \frac{\sqrt{h_1}}{\sqrt{h_2}} v_2 \sin\beta_2$$

And

$$\frac{v_1 \cos\beta_1}{r_1} = \frac{v_2 \cos\beta_2}{r_2}$$

$$\tan\beta_1 = \tan\beta_2 \quad \rightarrow \quad \beta_1 = \beta_2$$

But by proving the above way, we don't prove:

$$v^2 = v_1{}^2 + v_2{}^2 + 2v_1v_2$$

Or

$$v = v_1 + v_2$$

Which was the target and the sole purpose to balance the Tahir Gravitivistic Force of Gravitation between two bodies that they exert on each other under binary-star systems. So, some of the Tahirian Universal laws of the binary-star systems are as below:

$$\frac{v_1}{v_2} = \frac{r_1}{r_2} = \frac{\sqrt{h_1}}{\sqrt{h_2}}$$

Two bodies can't KEEP ON attracting each other unless their momentum is equal. Why! Because if their momenta are not the same how come their rate of change of momentum equals force, since the force exerted by one body on the second is equal in magnitude to the force exerted by second on the first. It's a mutual give and take. In other words, the magnitude of the momentum induced by one body into the other must be equal to the magnitude of the momentum induced by the other onto the first. *Let us call this principle as Tahir Universal Principal of Momentum of Binary Star Systems.* Amazing, isn't it! So, differentiating the equation below w.r.t. T we obtain

$$m_2v_2 = m_1v_1 = p$$

Or

$$\frac{v_1}{v_2} = \frac{m_2}{m_1}$$

$$m_2\frac{dv_2}{dT} = m_1\frac{dv_1}{dT} = \frac{dp}{dT} = \text{Force}$$

$$m_2\frac{dv_2}{dT} = m_1\frac{dv_1}{dT}$$

Also $v = v_1 + v_2$

Differentiating the above equation, we obtain as below:

$$\frac{dv}{dT} = \frac{dv_1}{dT} + \frac{dv_2}{dT}$$

Inserting one equation into the other, we obtain

$$m_2\frac{dv_2}{dT} = m_1\frac{dv_1}{dT}$$

$$m_2\frac{dv_2}{dT} = m_1\left(\frac{dv}{dT} - \frac{dv_2}{dT}\right)$$

$$m_2\frac{dv_2}{dT} + m_1\frac{dv_2}{dT} = m_1\frac{dv}{dT}$$

$$(m_2 + m_1)\frac{dv_2}{dT} = m_1\frac{dv}{dT}$$

$$\frac{dv_2}{dT} = \frac{m_1}{(m_2 + m_1)}\frac{dv}{dT}$$

&

$$\frac{dv_1}{dT} = \frac{m_2}{(m_2 + m_1)}\frac{dv}{dT}$$

Now I will repeat what I said before that the final hit in history can only be done by Tahir Standard Consciousness (TSC) and the leap just simply can't be made by the Algebra of AlKhawarizmi and the Calculus of Ibn ul Haishem! Algebra and Calculus can only take us close to the results and simply don't have the power to conclude. Concluding thus as a result now we obtain as below:

$$\frac{dv_2}{dT} = -(1 + 3t_2^2\frac{v_2^2}{s^2})\frac{Gm_1}{r^2}\sin\beta_2$$

$$\frac{dv_1}{dT} = -(1 + 3t_1^2\frac{v_1^2}{s^2})\frac{Gm_2}{r^2}\sin\beta_1$$

Where $t_2 = 1 + \frac{m_2}{m_1}$ & $t_1 = 1 + \frac{m_1}{m_2}$

$$\beta_1 = \beta_2 = \beta$$

$$\frac{d\beta_1}{dT} = -(1 + 3t_2^2\frac{v_2^2}{s^2})\frac{Gm_2}{r^2}\frac{\cos\beta_1}{v_1} + \frac{v_1\cos\beta_1}{r_1}$$

&

$$\frac{d\beta_2}{dT} = -(1 + 3t_2^2\frac{v_2^2}{s^2})\frac{Gm_1}{r^2}\frac{\cos\beta_2}{v_2} + \frac{v_2\cos\beta_2}{r_2}$$

$$\frac{dr_2}{dT} = v_2\sin\beta_2 \,\&\, \frac{dr_1}{dT} = v_1\sin\beta_1$$

$$h_2 = r_2v_2\cos\beta_2 \,\&\, h_1 = r_1v_1\cos\beta_1$$

$$(1 + 3t_1^2\frac{v_1^2}{s^2})\frac{Gm_2}{r^2}\cos\beta_1 = \left(\frac{v_1^2\cos^2\beta_1}{r_1} - \frac{d^2r_1}{dT^2}\right)\cos\beta_1$$

$$(1 + 3t_2^2\frac{v_2^2}{s^2})\frac{Gm_1}{r^2}\cos\beta_2 = \left(\frac{v_2^2\cos^2\beta_2}{r_2} - \frac{d^2r_2}{dT^2}\right)\cos\beta_2$$

Tahir Equations of Binary Star Systems

$$\frac{m_2}{m_1} = \frac{v_1}{v_2} = \frac{r_1}{r_2} = \frac{\sqrt{h_1}}{\sqrt{h_2}}$$

Where:

$$m = m_1 + m_2$$
$$v = v_1 + v_2$$
$$r = r_1 + r_2$$
$$\sqrt{h} = \sqrt{h_1} + \sqrt{h_2}$$

$$m_2\left(1 + \frac{m_1}{m_2}\right) = m$$

$$v_2\left(1 + \frac{m_2}{m_1}\right) = v$$

$$r_2\left(1 + \frac{m_2}{m_1}\right) = r$$

$$h_2\left(1 + \frac{m_2}{m_1}\right)^2 = h$$

The above equations completely describe the binary star-systems all over in the Tahirian Cosmos.

One very amazing thing to note is that the mystery of mass m_1 is solved by declaring it as just one of the parameters of motion under gravitation like the v_1, r_1, ρ_1. *There is nothing mysterious about mass anymore in history henceforward. Space, time and mass are interconnected, in the fabric of the gravitational force of the binary star system substituted in place of the centripetal acceleration.*

$$\frac{d\beta}{dT} = -\kappa\frac{Gm}{r^2}\frac{\cos\beta}{v} + \frac{v\cos\beta}{r}$$

$$\frac{dr}{dT} = v\sin\beta \ \& \ h = rv\cos\beta$$

$$\frac{d^2r}{dT^2} - \frac{v^2\cos^2\beta}{r} = -\frac{Gm}{r^2}\left\{1 + 3\frac{v^2}{s^2}\right\}$$

From $h = rv\cos\beta = (r_1 + r_2)(v_1 + v_2)\cos\beta$

We can easily deduce,

$$\sqrt{h} = \sqrt{h_1} + \sqrt{h_2}$$

by using the Tahirian equations of Tahirian Binary star systems as below as starting point

$$r_1v_1 = \frac{h_1}{h_2}r_2v_2$$

So, basically, what I have done is that I have been able to establish the equations of the system of attraction between two particles of matter or stars in terms of $m, r, v, , h$ and θ as below

$$\left\{1 + 3\frac{v^2}{s^2}\right\}\frac{Gm}{r^2} = \frac{v^2\cos^2\beta}{r} - \frac{d^2r}{dT^2}$$

$$\frac{dr}{dT} = v\sin\beta \ \& \ h = rv\cos\beta$$

$$\frac{d\beta}{dT} = -(1 + 3\frac{v^2}{s^2})\frac{Gm}{r^2}\frac{\cos\beta}{v} + \frac{v\cos\beta}{r}$$

$$\frac{dv}{dT} = -(1 + 3\frac{v^2}{s^2})\frac{Gm}{r^2}\sin\beta$$

$$(1 + 3\frac{v^2}{s^2})\frac{Gm}{r^2}\cos\beta = (\frac{v^2\cos^2\beta}{r} - \frac{d^2r}{dT^2})\cos\beta$$

$$\frac{d\theta}{dT} = \frac{h}{r^2}$$

$$\frac{d\beta}{d\theta} = 1 - (1 + 3\frac{v^2}{s^2})\frac{Gm}{v^2r}$$

$$d\theta = \frac{dr}{r\tan\beta}$$

Now, in order, to prove the above, we will use chain rule as below:

$$\frac{d\beta}{d\theta} = \frac{d\beta}{dT}\frac{dT}{d\theta} = (-(1 + 3\frac{v^2}{s^2})\frac{Gm}{r^2}\frac{\cos\beta}{v} + \frac{v\cos\beta}{r})(\frac{r^2}{h})$$

$$\frac{d\beta}{d\theta} = 1 - (1 + 3\frac{v^2}{s^2})\frac{Gm}{v^2r}$$

Once Tahir, *the Pure*, has established the equations of the system of two particles attracting each other, the next step is to split the system into corresponding equations of the two separate particles with TCR as the Tahir Centre of Rotation as below:

$$\frac{d^2r_2}{dT^2} - \frac{v_2^2\cos^2\beta_2}{r_2} = -\frac{Gm_1}{r^2}\left\{1 + 3t_2^2\frac{v_2^2}{s^2}\right\}$$

$$\frac{dr_2}{dT} = v_2\sin\beta_2 \ \& \ h_2 = r_2v_2\cos\beta_2$$

$$\frac{d\beta_2}{dT} = -\left\{1 + 3t_2^2\frac{v_2^2}{s^2}\right\}\frac{Gm_1}{r^2}\frac{\cos\beta_2}{v_2} + \frac{v_2\cos\beta_2}{r_2}$$

$$\frac{dv_2}{dT} = -(1 + 3t_2^2 \frac{v_2^2}{s^2}) \frac{Gm_1}{r^2} \sin\beta_2$$

$$\frac{d\theta_2}{dT} = \frac{h_2}{r_2^2}$$

$$\frac{d\beta_2}{d\theta_2} = 1 - (1 + 3t_2^2 \frac{v_2^2}{s^2}) \frac{Gm_1}{t_2^2 v_2^2 r_2}$$

$$d\theta_2 = \frac{dr_2}{r_2 \tan\beta_2}$$

Similar equations can be found for the other particle with suffix 1. Both particles are rotating with respect to the Tahir Centre of Rotation which is considered fixed like origin in Omar Khayyam's Coordinates x and y for the Binary Particles motion.

The equations connecting the system and particles are the following and as went above,

$$\frac{m_2}{m_1} = \frac{v_1}{v_2} = \frac{r_1}{r_2} = \frac{\sqrt{h_1}}{\sqrt{h_2}}$$

Where:

$$m = m_1 + m_2$$

$$v = v_1 + v_2$$

$$r = r_1 + r_2$$

$$\sqrt{h} = \sqrt{h_1} + \sqrt{h_2}$$

Or

$$m_2 \left(1 + \frac{m_1}{m_2}\right) = m$$

$$v_2 \left(1 + \frac{m_2}{m_1}\right) = v$$

$$r_2 \left(1 + \frac{m_2}{m_1}\right) = r$$

$$h_2 \left(1 + \frac{m_2}{m_1}\right)^2 = h$$

The common parameters connecting the system and the two particles are the following:

$$\beta_1 = \beta_2 = \beta$$
$$\theta_1 = \theta_2 = \theta$$

The last equation above can be slightly difficultly understood by equating $\frac{d\beta}{d\theta}$ equal to zero and the resulting substitution of $m_1 = m_2$. Care should be exercised to keep the system axes

unrotated on one mass for relative values of θ. It constitutes motion of two particles of same mass in a circular orbit. If $\frac{d\beta}{d\theta} \neq 0$ and the masses are equal, then it constitutes elliptical-like (but not elliptical static orbits) motion of two masses in elliptical like precessing orbits. On the contrary, there is absolutely no gravitivistic precession of the perihelion of one mass or the other in a circular orbit just discussed above.

Where $t_2 = 1 + \frac{m_2}{m_1}$

The Crux of Tahirian Cosmology is that when the two masses are unequal in magnitude then the accelerations of the particles induced in each other are different even though the magnitude of the gravitational force induced on each other is the same. But when masses are equal both the induced quantities are the same.

Now if we consider the equations as below,

$$\frac{d^2 r_2}{dT^2} - \frac{v_2^2 \cos^2\beta_2}{r_2} = -\frac{Gm_1}{r^2}\left\{1 + 3t_2^2 \frac{v_2^2}{s^2}\right\}$$

$$\frac{d^2 r_1}{dT^2} - \frac{v_1^2 \cos^2\beta_2}{r_1} = -\frac{Gm_2}{r^2}\left\{1 + 3t_1^2 \frac{v_1^2}{s^2}\right\}$$

we can prove another Great Formula concerning Tahirian Cosmology by multiplying the above equations by m_1 & m_2.

$$m_2 \frac{d^2 r_2}{dT^2} - \frac{m_2 v_2^2 \cos^2\beta_2}{r_2} = -\frac{Gm_1 m_2}{r^2}\left\{1 + 3t_2^2 \frac{v_2^2}{s^2}\right\}$$

$$m_1 \frac{d^2 r_1}{dT^2} - \frac{m_1 v_1^2 \cos^2\beta_1}{r_1} = -\frac{Gm_2 m_1}{r^2}\left\{1 + 3t_1^2 \frac{v_1^2}{s^2}\right\}$$

Subtracting one from the other we finally obtain as below:

$$m_1 \frac{d^2 r_1}{dT^2} - m_2 \frac{d^2 r_2}{dT^2} = \frac{m_1 v_1^2 \cos^2\beta_2}{r_1} - \frac{m_2 v_2^2 \cos^2\beta_1}{r_2}$$

We can easily prove that both the RHS and the LHS of the above equation reduce to zero. Let us see how.

$$\frac{dr_1}{dT} = v_1 \sin\beta_1$$
$$\frac{dr_2}{dT} = v_2 \sin\beta_2$$

Since in Tahirian Cosmology $\beta_1 = \beta_2$. Substituting one value into the other, we obtain as below.

$$v_2 \frac{dr_1}{dT} = v_1 \frac{dr_2}{dT}$$

$$\frac{v_1}{v_2} = \frac{m_2}{m_1}$$

$$m_1 \frac{dr_1}{dT} = m_2 \frac{dr_2}{dT}$$

Differentiating w.r.t. T once again, we obtain as below:

$$m_1 \frac{d^2 r_1}{dT^2} = m_2 \frac{d^2 r_2}{dT^2}$$

Substituting the above equation in the equation below we obtain an important result regarding binaries.

$$m_1 \frac{d^2 r_1}{dT^2} - m_2 \frac{d^2 r_2}{dT^2} = \frac{m_1 v_1^2 \cos^2 \beta_2}{r_1} - \frac{m_2 v_2^2 \cos^2 \beta_1}{r_2}$$

So, we obtain another result as below:

$$\frac{m_1 v_1^2 \cos^2 \beta_2}{r_1} = \frac{m_2 v_2^2 \cos^2 \beta_1}{r_2}$$

Now since $\beta_1 = \beta_2$, we obtain eventually as below:

$$\frac{m_1 v_1^2}{r_1} = \frac{m_2 v_2^2}{r_2}$$

Or

$$\frac{m_1 v_1^2 r_1^2 \cos^2 \beta_1}{r_1^3} = \frac{m_2 v_2^2 r_2^2 \cos^2 \beta_2}{r_2^3}$$

$$\frac{m_1 h_1^2}{r_1^3} = \frac{m_2 h_2^2}{r_2^3}$$

Substituting the above results in the gravitivistic proportion of the equation below

$$m_1 \frac{d^2 r_1}{dT^2} - \frac{m_1 v_1^2 \cos^2 \beta_1}{r_1} = -\frac{Gm_2 m_1}{r^2} \left\{ 1 + 3t_1^2 \frac{v_1^2}{s^2} \right\}$$

$$m_1 \frac{d^2 r_1}{dT^2} - \frac{m_1 v_1^2 \cos^2 \beta_1}{r_1} = -\frac{Gm_2 m_1}{r^2} \left\{ 1 + 3t_1^2 \frac{v_1^2}{s^2} \right\}$$

$$m_1 \frac{d^2 r_1}{dT^2} - \frac{m_1 v_1^2 \cos^2 \beta_1}{r_1} = -\frac{Gm_2 m_1}{r^2} - 3t_1^2 \frac{v_1^2}{s^2} \frac{Gm_2 m_1}{r^2}$$

$$m_1 \frac{d^2 r_1}{dT^2} - \frac{m_1 v_1^2 \cos^2 \beta_1}{r_1} = -\frac{Gm_2 m_1}{r^2} - 3t_1^2 \frac{v_1^2}{r^2} \frac{Gm_2 m_1}{s^2}$$

$$r_1 \left(1 + \frac{m_1}{m_2} \right) = r$$

Where $t_1 = 1 + \frac{m_1}{m_2}$

$$r_1 t_1 = r$$

$$m_1 \frac{d^2 r_1}{dT^2} - \frac{m_1 v_1^2 \cos^2 \beta_1}{r_1} = -\frac{Gm_2 m_1}{r^2} - 3 \frac{v_1^2}{r_1^2} \frac{Gm_2 m_1}{s^2}$$

$$m_1 \frac{d^2 r_1}{dT^2} - \frac{m_1 v_1^2 \cos^2 \beta_1}{r_1} = -\frac{Gm_2 m_1}{r^2} - 3Q \frac{Gm_2 m_1}{s^2}$$

$$m_1 \frac{d^2 r_1}{dT^2} - \frac{m_1 v_1^2 \cos^2 \beta_1}{r_1} = -\frac{Gm_2 m_1}{r^2} - 3Q \frac{Gm_2 m_1}{s^2}$$

$$m_1 \frac{d^2 r_1}{dT^2} - \frac{m_1 h_1^2}{r_1^3} = -\frac{Gm_2 m_1}{r^2} - 3Q \frac{Gm_2 m_1}{s^2}$$

$$m_1 \frac{d^2 r_1}{dT^2} - \frac{S}{r_1^3} = \frac{R}{r_1^2} - 3Q \frac{Gm_2 m_1}{s^2}$$

$$m_1 \frac{d^2 r_1}{dT^2} = \frac{S}{r_1^3} + \frac{R}{r_1^2} - 3Q \frac{Gm_2 m_1}{s^2}$$

$$S = m_1 h_1^2$$

$$R = -\frac{Gm_2 m_1}{t_1^2}$$

$$Q = \frac{v_1^2}{r_1^2}$$

Differentiating the above equation w.r.t. T, we obtain as below:

$$m_1 \frac{d^3 r_1}{dT^3} = -3 \frac{Gm_2 m_1}{s^2} \frac{dQ}{dT} - 2R r_1^{-3} \frac{dr_1}{dT} - 3S r_1^{-4} \frac{dr_1}{dT}$$

Differentiating the above equation w.r.t. T again, we obtain as below:

$$m_1 \frac{d^m}{dT^m} \left[\frac{d^3 r_1}{dT^3} \right] = -3 \frac{Gm_2 m_1}{s^2} \frac{d^m Q}{dT^m} - 2R \frac{d^m}{dT^m} [r_1^{-3} \frac{dr_1}{dT}] - 3S \frac{d^m}{dT^m} [r_1^{-4} \frac{dr_1}{dT}]$$

The purpose of all the above calculations of Tahir is to write the Tahir-Ibn-Sina-Abul-Barkat Al-Baghdadi-Ibne-Bajjaj-noton-anstan equation in a form whose nth derivatives w.r.t. $r_{1\,\&}$ T can be easily deduced as below:

I first found the nth derivative with n as positive and then I made the change that was meant to be made due to the negative exponent n.

$$\frac{d^m}{dT^m} \left[r^n \frac{dr}{dT} \right] = \sum_{k=0}^{m} \frac{m!}{k!(m-k)!} \frac{n!}{(n-m+k)!} r^{n-m+k} \frac{d^k}{dT^k} [\frac{dr}{dT}]$$

Where $n = 1, 2, 3, \ldots \infty$ **&**
$k = 0, 1, 2, 3 \ldots \ldots m$

The above can be easily transformed to the required form with negative exponents, as in our case, as below:

$$\frac{d^m}{dT^m} \left[r^{-n} \frac{dr}{dT} \right] = \sum_{k=0}^{m} (-1)^{m-k} \frac{m!}{k!(m-k)!} \frac{(n+m-k-1)!}{(n-1)!} r^{-n-m+k} \frac{d^k}{dT^k} [\frac{dr}{dT}]$$

Where $n = 1, 2, 3, \ldots \infty$ **&**
$k = 0, 1, 2, 3 \ldots \ldots m$

It's amazing because no one in history ever thought that

Natural Theory of Relativity, Inertia, Gravitation & Gravitivity

anyone would ever even think of finding the nth derivative of noton-anstan incomplete equation of gravitation!

I have utilized Tahir Convention above as below:

$$\frac{d^0}{dT^0}\left[\frac{d^1 r}{dT^1}\right] = \frac{d^1 r}{dT^1}$$

$$\frac{d^1}{d^1}\left[\frac{d^1 r}{dT^1}\right] = \frac{d^2 r}{dT^2}$$

$$\frac{d^2}{dT^2}\left[\frac{d^1 r}{dT^1}\right] = \frac{d^3 r}{dT^3}$$

$$\vdots$$

$$\frac{d^f}{dT^f}\left[\frac{d^1 r}{dT^1}\right] = \frac{d^{f+1} r}{dT^{f+1}}$$

I am not concerned with any derivative at n = 0 but those who would be, they must consider the condition of m = k as well along with it.

Higher order derivatives w.r.t. r can also be found for the equation below as

$$m_1 \frac{d^2 r_1}{dT^2} = -3Q\frac{Gm_2 m_1}{s^2} + \frac{R}{r_1^2} + \frac{S}{r_1^3}$$

$$m_1 \frac{d^m}{dr^m}\left[\frac{d^2 r_1}{dT^2}\right] = -3\frac{Gm_2 m_1}{s^2}\frac{d^m Q}{dT^m} + R\frac{d^m}{dr^m}[r_1^{-2}] + S\frac{d^m}{dr^m}[r_1^{-3}]$$

$$\frac{d^m}{dr^m}[r_1^{-n}] = (-1)^m \frac{(n+m-1)!}{(n-1)!} r_1^{-n-m}$$

Where m = 0, 1, 2, 3, ... ∞

Since $\frac{dQ}{dT} \neq 0$ in the orbit, therefore we can further expand Q in terms of the variable r or r_{22} as below and as will come next.

$$v_{22}^2 = \left(v_{21}^2 - \frac{2Gm_1}{t_2^2 r_{21}} + \frac{6G^2 m_1^2}{t_2^2 s^2 r_{21}^2} - \frac{6Gm_1 v_{21}^2}{s^2 r_{21}}\right) + \left(\frac{2Gm_1}{t_2^2}\right.$$
$$\left. - \frac{12G^2 m_1^2}{t_2^2 s^2 r_{21}} + \frac{6Gm_1 v_{21}^2}{s^2}\right)\frac{1}{r_{22}} + \frac{6G^2 m_1^2}{t_2^2 s^2}\frac{1}{r_{22}^2}$$
$$+ \cdots$$

$$Q = \frac{v_{22}^2}{r_{22}^2} = \left(v_{21}^2 - \frac{2Gm_1}{t_2^2 r_{21}} + \frac{6G^2 m_1^2}{t_2^2 s^2 r_{21}^2} - \frac{6Gm_1 v_{21}^2}{s^2 r_{21}}\right)\frac{1}{r_{22}^2}$$
$$+ \left(\frac{2Gm_1}{t_2^2} - \frac{12G^2 m_1^2}{t_2^2 s^2 r_{21}} + \frac{6Gm_1 v_{21}^2}{s^2}\right)\frac{1}{r_{22}^3}$$
$$+ \frac{6G^2 m_1^2}{t_2^2 s^2}\frac{1}{r_{22}^4} + \cdots$$

$$Q = \frac{v_{22}^2}{r_{22}^2} = \frac{c}{r_{22}^2} + \frac{b}{r_{22}^3} + \frac{a}{r_{22}^4} + \cdots$$

Or

$$Q = \frac{v_1^2}{r_1^2} = \frac{c}{r_1^2} + \frac{b}{r_1^3} + \frac{a}{r_1^4} + \cdots$$

Subscripts can be interchanged without discomfort.

This will give the exact m^{th} derivative w.r.t. T of the complete equation of binary star systems up to the fourth power of s, i.e., s^4.

So, summing up the story started by Ibn-e-Sina, the father of Inertia and Gravitation, we have for inertia:

$$r_1 v_1 \cos\beta_1 = h_1$$

$$r_1 = \frac{h_1}{v_1 \cos\beta_1}$$

$$\frac{dr_1}{dT} = \frac{d}{dT}\left[\frac{h_1}{v_1 \cos\beta_1}\right] = v_1 \sin\beta_1$$

$$\frac{d^2 r_1}{dT^2} = \frac{v_1^2 \cos^2\beta_1}{r_1}$$

$$\frac{dr_1}{dT} = \pm i v_1 \cos\beta_1$$

$$\frac{d^n}{dr_1^n}\left[\frac{d^2 r_1}{dT^2}\right] = \frac{(-1)^n}{2!}(n+2)!\frac{h_1^2}{r_1^{n+3}}$$

Where n = 0, 1, 2, 3, ... ∞

$$\frac{dv_1}{dT} = 0 \implies v_1 = \textbf{Constant (Uniform speed)}$$

$$V_1' = V_1 \sin\beta_1 - \frac{dr_1}{dT}$$

$$V_1' = V_1 \sin\beta_1 - \pm i v_1 \cos\beta_1 \quad (= 0 \text{ for inertial motion})$$

$$V_1' = v_1 e^{\mp i(\frac{\pi}{2}-\beta_1)} = 0$$

$$v_1 e^{\mp i(\frac{\pi}{2}-\beta_1)} = 0 \ \& \ \omega = 0 \text{ as well}$$
(Equations of the Supreme Throne)

$$v_1 = 0 \ \& \ \omega = 0 \quad \textbf{Qayyam (Absolute Rest)}$$

Both the motions, translational and spinning, of the Throne are void!

Al-Qayyum – ALLAH Kareem – the Supreme – is the only One who has qayyam – Absolute rest.

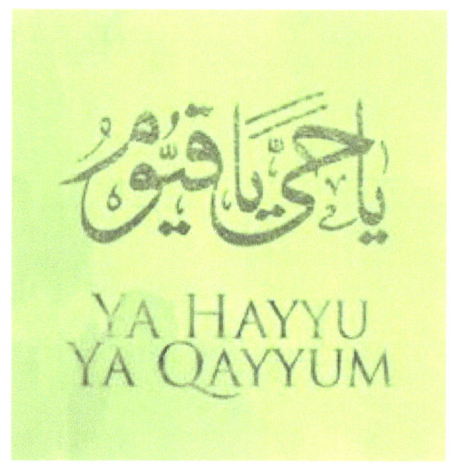

YA HAYYU
YA QAYYUM

The Ever Living and the Possessor of the Throne of Absolute Rest

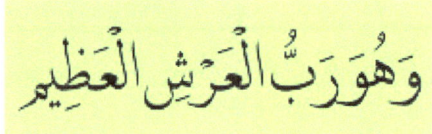

…and HE is the Lord of the Supreme Throne

I bear testimony that there is no deity/god/creator worthy of worship except the *Lord* of Muhammed (the blessed Son of the blessed mother - Amina Bint Wahb) *ALLAH KAREEM THE SUPREME* and *HE* only has got the Absolute Rest. *Ya Hayyu, Ya Qayyum.*

$$\frac{2}{1!} + \frac{4}{3!} + \frac{6}{5!} + \cdots \infty = \frac{1}{0!} + \frac{3}{2!} + \frac{5}{4!} + \frac{7}{6!} + \cdots \infty$$

$$1 = \frac{0}{1!} + \frac{1}{2!} + \frac{2}{3!} + \cdots \infty = Ahad$$

$$e = \frac{2}{1!} + \frac{4}{3!} + \frac{6}{5!} + \cdots = \frac{1}{0!} + \frac{3}{2!} + \frac{5}{4!} + \frac{7}{6!} + \cdots \infty$$

One is an odd number. It's the symbol for the ONENESS of ALLAH Kareem – the Supreme.

Using Tahir Series expansion of functions as below

$$f(x) = f(0) + xf'(x) - \frac{x^2}{2!}f''(x) + \frac{x^3}{3!}f'''(x) - \cdots$$

Let $f(x) = xe^x$
$$f^n(x) = (x + n)e^x$$
$$n = 0, 1, 2, 3, \ldots \infty$$

$$f^0(x) = (x + 0)e^x = f(x)$$
$$f^1(x) = (x + 1)e^x$$
$$\vdots$$

$$f^n(x) = (x + n)e^x$$

Tahir Convention is used again as below:

$$xe^x = 0 + x(x + 1)e^x - \frac{x^2}{2!}(x + 2)e^x + \frac{x^3}{3!}(x + 3)e^x$$

$$1 = \frac{(x + 1)}{1!} - \frac{(x^2 + 2x)}{2!} + \frac{x^3 + 3x^2}{3!} - \frac{x^4 + 4x^3}{4!} - \cdots \infty$$

At $x = +1$ & -1 I obtain the above Tahir Series Expansions of ONE.

A more commendable form of the inertial equations just went before can be written as below:

$$r_1 = \frac{h_1}{v_1 \sin\theta_1}$$

$$\frac{dr_1}{dT} = \pm iv_1 \sin\theta_1$$

$$\frac{dr_1}{dT} = v_1 \cos\theta_1$$

$$\frac{d^2 r_1}{dT^2} = \frac{v_1^2 \sin^2\theta_1}{r_1}$$

$$\frac{d^n}{dr_1^n}\left[\frac{d^2 r_1}{dT^2}\right] = \frac{(-1)^n}{2!}(n + 2)! \frac{h_1^2}{r_1^{n+3}}$$

Where $n = 0, 1, 2, 3, \ldots \infty$

The formula for the nth derivative w.r.t. T has already been got for the gravitivistic case.

$$\frac{dv_1}{dT} = 0 \quad \Rightarrow \quad v_1 = \textbf{Constant (Uniform speed)}$$

$$v_1 e^{\mp i\theta_1} = 0$$

(Equation of the Supreme Throne)

$v_1 = 0$ Qayyam (Absolute Rest)

$$\beta = \frac{\pi}{2} - \theta$$
$$\theta = \frac{\pi}{2} - \beta$$

A body continues in its state of rest(without spinning), i.e., V = 0 & ω = 0 unless compelled by some external force/torque to act otherwise.

Where one arm of the angle θ is parallel to both velocities v_1 and v_2 and the other arms go to the moving objects subscripted 1 and 2 above. Both the objects/stars move under Tahir's Command in physics called as Tahir Hypothetical Inertial Screening (THIS) whereby both stars are screened from the gravitational effects of each-other and they eventually start moving in parallel lines in opposite directions. THIS cannot be compared with the naive concept of inertial catastrophe in which the Sun vaporizes. Diagrams are left as an aftermath of the extreme struggle of understanding Tahir. Both arms of the angle θ intersect at TCR.

Now please let me find the Tahirian Circular Orbital Speed by using one of the Laws pertaining to Tahirian Cosmology as below:

$$\frac{d\beta}{d\theta} = 1 - \left(1 + 3\frac{v^2}{s^2}\right)\frac{Gm}{v^2r} = 0$$

For circular orbital motion $\frac{d\beta}{d\theta}$ (or $\frac{d\beta}{dT}$) $= 0$. Replacing r by r_2 or converting the equation of the system of binaries into its one binary component we obtain

$$1 - \left(1 + 3\frac{v_2^2}{s^2}t_2^2\right)\frac{Gm}{v_2^2t_2^2r}$$

$$\frac{v_2^2}{r_2t_2} = \left\{1 + 3t_2^2\frac{v_2^2}{s^2}\right\}\frac{Gm_1t_2}{t_2^2r_2^2t_2^2}$$

$$\frac{v_2^2}{r_2} - \left\{3t_2^2\frac{v_2^2}{s^2}\right\}\frac{Gm_1}{r_2^2t_2^2} = \frac{Gm_1}{r_2^2t_2^2}$$

$$v_2^2 - \left\{3t_2^2\frac{v_2^2}{s^2}\right\}\frac{Gm_1}{r_2t_2^2} = \frac{Gm_1}{r_2t_2^2}$$

$$v_2^2\left(1 - 3\frac{Gm_1}{r_2s^2}\right) = \frac{Gm_1}{t_2^2r_2}$$

$$v_2^2 = \frac{Gm_1}{t_2^2r_2\left(1 - 3\frac{Gm_1}{r_2s^2}\right)}$$

Let $R_2 = r_2\left(1 - 3\frac{Gm_1}{r_2s^2}\right)$

$$v_2^2 = \frac{Gm_1}{t_2^2R_2}$$

Now let us find the value of the speed of two stars of equal mass moving in the same circular orbit, if they can move, as below

$$\frac{d\beta}{d\theta} = 1 - \left(1 + 3\frac{v^2}{s^2}\right)\frac{Gm}{v^2r} = 0$$

$$\left(1 + 3\frac{v^2}{s^2}\right)\frac{G(m_1 + m_2)}{(v_1 + v_2)^2(r_1 + r_2)} = 1$$

$$\left(1 + 3\frac{v^2}{s^2}\right)\frac{Gm_1(1 + \frac{m_2}{m_1})}{v_2^2\left(1 + \frac{v_1}{v_2}\right)^2 r_2(1 + \frac{r_1}{r_2})} = 1$$

Now we know from Tahirian Cosmology that the following holds true as went before:

$$\frac{m_2}{m_1} = \frac{v_1}{v_2} = \frac{r_1}{r_2} = \frac{\sqrt{h_1}}{\sqrt{h_2}}$$

So, finally we obtain as below:

$$\left(1 + 3\frac{(2v_2)^2}{s^2}\right)\frac{Gm_1}{v_2^2\left(1 + \frac{v_1}{v_2}\right)^2 r_2} = 1$$

$$\left(1 + 12\frac{v_2^2}{s^2}\right)\frac{Gm_1}{v_2^2\left(1 + \frac{v_1}{v_2}\right)^2 r_2} = 1$$

$$\left(1 + 12\frac{v_2^2}{s^2}\right)\frac{Gm_1}{4v_2^2r_2} = 1$$

$$v_2^2 = \left(1 + 12\frac{v_2^2}{s^2}\right)\frac{Gm_1}{4r_2}$$

$$v_2^2 = \frac{Gm_1}{4r_2} + 3\frac{v_2^2}{s^2}\frac{Gm_1}{r_2}$$

$$v_2^2 - 3\frac{v_2^2}{s^2}\frac{Gm_1}{r_2} = \frac{Gm_1}{4r_2}$$

$$v_2^2 = \frac{1}{(1 - 3\frac{Gm_1}{r_2s^2})}\frac{Gm_1}{4r_2}$$

$$v_2 = \sqrt{\frac{Gm_1}{4r_2(1 - 3\frac{Gm_1}{r_2s^2})}}$$

Let us derive the work energy theorem. It is basically a sin to call it work energy theorem. May ALLAH Kareem – the SUPREME forgive Tahir's sins, big and small, hidden and disclosed, intentional and unintentional and grant him forgiveness and let him enter Jannat-ul-Firdous with utmost welcome. Ameen! Ya Rubbal Alameen. Let us call it Tahirian Theorem

$$\frac{dv_2}{dT} = -(1 + 3t_2^2\frac{v_2^2}{s^2})\frac{Gm_1}{r^2}\sin\beta_2 \quad \& \quad \frac{dr_2}{dT} = v_2\sin\beta_2$$

$$\int_{v_{21}}^{v_{22}} \frac{v_2 dv_2}{\left\{1 + 3t_2^2 \frac{v_2^2}{s^2}\right\}} = -\int_{r_{21}}^{r_{22}} \frac{Gm_1}{r^2} dr_2$$

$$\frac{s^2}{6t_2^2} \int_{v_{21}}^{v_{22}} \frac{d\left(1 + 3t_2^2 \frac{v_2^2}{s^2}\right)}{\left\{1 + 3t_2^2 \frac{v_2^2}{s^2}\right\}} = -\int_{r_{21}}^{r_{22}} \frac{Gm_1}{r^2} dr_2$$

$$t_2 = 1 + \frac{m_2}{m_1}$$

$$r^2 = r_2^2 \left(1 + \frac{m_2}{m_1}\right)^2 = r_2^2 t_2^2$$

We obtain as below:

$$\frac{s^2}{6} \int_{v_{21}}^{v_{22}} \frac{d\left(1 + 3t_2^2 \frac{v_2^2}{s^2}\right)}{\left\{1 + 3t_2^2 \frac{v_2^2}{s^2}\right\}} = -\int_{r_{21}}^{r_{22}} \frac{Gm_1}{r_2^2} dr_2$$

$$\int_{v_{21}}^{v_{22}} \frac{d\left(1 + 3t_2^2 \frac{v_2^2}{s^2}\right)}{\left\{1 + 3t_2^2 \frac{v_2^2}{s^2}\right\}} = \frac{6Gm_1}{s^2}\left(\frac{1}{r_{22}} - \frac{1}{r_{21}}\right)$$

$$\ln\left(\frac{1 + 3t_2^2 \frac{v_{22}^2}{s^2}}{1 + 3t_2^2 \frac{v_{21}^2}{s^2}}\right) = \frac{6Gm_1}{s^2}\left(\frac{1}{r_{22}} - \frac{1}{r_{21}}\right)$$

$$1 + 3t_2^2 \frac{v_{22}^2}{s^2} = \left(1 + 3t_2^2 \frac{v_{21}^2}{s^2}\right) e^{\left[\frac{6Gm_1}{s^2}\left(\frac{1}{r_{22}} - \frac{1}{r_{21}}\right)\right]}$$

$$v_{22}^2 = \frac{s^2}{3t_2^2}\left\{\left(1 + 3t_2^2 \frac{v_{21}^2}{s^2}\right) e^{\left[\frac{6Gm_1}{s^2}\left(\frac{1}{r_{22}} - \frac{1}{r_{21}}\right)\right]} - 1\right\}$$

By changing subscripts, we can obtain a similar equation for the other particle as below:

$$v_{12}^2 = \frac{s^2}{3t_1^2}\left\{\left(1 + 3t_1^2 \frac{v_{11}^2}{s^2}\right) e^{\left[\frac{6Gm_2}{s^2}\left(\frac{1}{r_{12}} - \frac{1}{r_{11}}\right)\right]} - 1\right\}$$

5. Projectile and planetary motion

Now let us find the time of flight of the projectile. Using Tahir's Fundamental Field Equation as went before we have:

$$\frac{dr}{dT} = v\sin\beta$$

$$dT = \frac{dr}{v\sin\beta}$$

$$dT = \frac{dr}{\sqrt{v_{22}^2 - \frac{h^2}{r_{22}^2}}}$$

$$T = 2\int_{R}^{r_0} \frac{dr}{\sqrt{v_{22}^2 - \frac{h^2}{r_{22}^2}}}$$

$$T = -2\int_{R}^{r_0} \frac{du_{22}}{u_{22}\sqrt{v_{22}^2 - h_2^2 u_{22}^2}}$$

Where r_0 is the maximum radial distance reached by the projectile from the Centre of the gravitating mass or simply Earth.

$$v_{22}^2 - h_2^2 u_{22}^2 \approx a u_{22}^2 + b u_{22} + c$$

$$a = \frac{6G^2 m_1^2}{t_2^2 s^2} - h_2^2$$

$$b = \frac{2Gm_1}{t_2^2} - \frac{12G^2 m_1^2}{t_2^2 s^2 r_{21}} + \frac{6Gm_1 v_{21}^2}{s^2}$$

$$c = v_{21}^2 - \frac{2Gm_1}{t_2^2 r_{21}} + \frac{6G^2 m_1^2}{t_2^2 s^2 r_{21}^2} - \frac{6Gm_1 v_{21}^2}{s^2 r_{21}}$$

Where from [3]

$$\int \frac{dx}{x^2\sqrt{R}} = -\frac{\sqrt{R}}{cx} - \frac{b}{2c\sqrt{-c}} \arcsin\frac{2c + bx}{x\sqrt{b^2 - 4ac}} \quad \{c < 0, \ \Delta < 0\}$$

$$\int \frac{dx}{x^2\sqrt{R}} = -\frac{\sqrt{R}}{cx} - \frac{b}{2c\sqrt{-c}} \arctan\frac{2c + bx}{2\sqrt{-c}\sqrt{R}} \quad \{c < 0\}$$

And

$$R = ax^2 + bx + c \quad \& \quad \Delta = 4ac - b^2$$

$$\therefore$$

$$T = 2\left|\left[\frac{\sqrt{R}}{cu} + \frac{b}{2c\sqrt{-c}} \arcsin\frac{2c + bu}{u\sqrt{b^2 - 4ac}}\right]\right|_{\frac{1}{R}}^{\frac{1}{r_0}} \quad \{c < 0, \ \Delta < 0\}$$

$$T = 2\left|\left[\frac{\sqrt{R}}{cu} + \frac{b}{2c\sqrt{-c}} \arctan\frac{2c + bu}{2\sqrt{-c}\sqrt{R}}\right]\right|_{\frac{1}{R}}^{\frac{1}{r_0}} \quad \{c < 0\}$$

In order, to find what I call as Proper Time T', we simply need to multiply dT by the factor $1 - \frac{v^2}{c^2}$ as below:

$$dT = \frac{dr}{v\sin\beta}$$

$$dT\left(1 - \frac{v^2}{c^2}\right) = \frac{dr}{v\sin\beta}\left(1 - \frac{v^2}{c^2}\right)$$

Natural Theory of Relativity, Inertia, Gravitation & Gravitivity

$$dT' = \frac{dr}{v\sin\beta}(1 - \frac{v^2}{c^2})$$

Integrating both sides, we obtain

$$T' = 2\left|(1 - \frac{v^2}{c^2})[\frac{\sqrt{R}}{cu} + \frac{b}{2c\sqrt{-c}}\arcsin\frac{2c+bu}{u\sqrt{b^2-4ac}}]\right|_{\frac{1}{R}}^{\frac{1}{r_0}} \quad \{c < 0, \ \Delta < 0\}$$

$$T' = 2\left|(1 - \frac{v^2}{c^2})[\frac{\sqrt{R}}{cu} + \frac{b}{2c\sqrt{-c}}\arctan\frac{2c+bu}{2\sqrt{-c}\sqrt{R}}]\right|_{\frac{1}{R}}^{\frac{1}{r_0}} \quad \{c < 0\}$$

Similarly, we can obtain an expression for the Tahirian Time t as below:

$$t = 2\left|(1 - 2\frac{v^2}{c^2})[\frac{\sqrt{R}}{cu} + \frac{b}{2c\sqrt{-c}}\arcsin\frac{2c+bu}{u\sqrt{b^2-4ac}}]\right|_{\frac{1}{R}}^{\frac{1}{r_0}} \quad \{c < 0, \ \Delta < 0\}$$

$$t = 2\left|(1 - 2\frac{v^2}{c^2})[\frac{\sqrt{R}}{cu} + \frac{b}{2c\sqrt{-c}}\arctan\frac{2c+bu}{2\sqrt{-c}\sqrt{R}}]\right|_{\frac{1}{R}}^{\frac{1}{r_0}} \quad \{c < 0\}$$

The factor $1 - \frac{v^2}{c^2}$ must be taken out of the integral sign and SHOULDN'T be integrated along with the rest of the TFE. However, it will be substituted with the values of v at the limits of the integral together with the result got after integrating the TFE. This factor belongs to Tahir Inertial Equation TIE and not to TFE! Moreover, the differentiation and integration have already been done once before FOREVER!

Above all, my entire theory satisfies the polar metric as below:

$$dS^2 = dr^2 + r^2 d\theta^2 = v^2\sin^2\beta dT^2 + v^2\cos^2\beta dT^2 = v^2 dT^2$$

Where

$$\frac{dr}{dT} = v\sin\beta \ \& \ v\cos\beta = r\frac{d\theta}{dT}$$

These two formulae above remained hid from the eyes of people for a bit over 300 years till the Advent-*of*–Tahir. They are some of the *Field Equations of Tahir*. When Tahir says Field Equations then he doesn't mean anything other than the FACT that they are equally true for both inertia and gravitation. Hence true for all types of Fields!

Dividing one with the other above we find another Field-Equation of Tahir as below:

$$\frac{dr}{d\theta} = r\tan\beta$$

We get:

$$d\theta = \frac{dr}{r\tan\beta}$$

$$\theta = \int_{r_1}^{r_2} \frac{dr}{r\tan\beta} \quad \text{One of the Field Equations of Tahir}$$

We will see that the above Tahir Field Equation TFE is equally true for inertia and gravitation. That is why I called such an equation as Tahir Field Equation (TFE). For inertia, it surrenders to Tahir's Pen as below in the form of an equation of a straight line.

$$r_1 = \frac{h_1}{v_1\sin\theta_1}$$

Care should be exercised to express $\tan\beta$ as $-\sqrt{\sec^2\beta - 1}$

For Tahir-Alien non-gravitivistic and Tahirian Gravitivistic solutions see other parts of my Theory. I will derive another Tahirian Formulae essential to describe the Tahirian Gravitation as below:

Using the formulae already derived, we have:

$$\frac{d\theta}{dT} = \frac{h}{r^2}$$

&

$$\frac{d\beta}{dT} = -\kappa\frac{GM}{r^2}\frac{\cos\beta}{v} + \frac{v\cos\beta}{r}$$

Now we will use chain rule as below:

$$\frac{d\beta}{d\theta} = \frac{d\beta}{dT}\frac{dT}{d\theta} = (-\kappa\frac{GM}{r^2}\frac{\cos\beta}{v} + \frac{v\cos\beta}{r})(\frac{r^2}{h})$$

$$\frac{d\beta}{d\theta} = 1 - \kappa\frac{Gm}{rv^2}$$

There is another way to prove it which I will show later if time permits. These are the Mathematical principles of Natural Philosophy contemplated by Tahir.

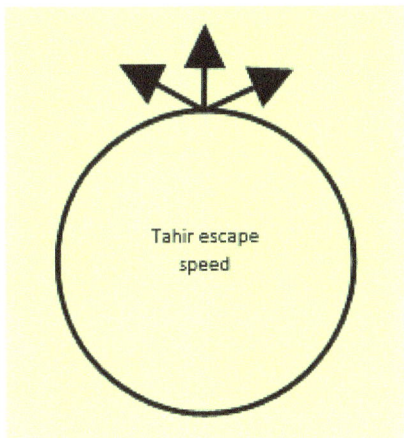

Tahir escape speed

Figure 20. *Tahir Escape Speed.*

Now the hands of the electronic print media would not be shaky to print the escape speed at an incline value of the angle other than 90° after reading the Ibn-e-Sina-Al-Kashi-Ibn-Muadh-Ibn-Rusdh-Alkhazin-Tahir gravitation and inertia. For those who praise the alien unnecessarily, let me tell them that it would not have been difficult for the Eye-*of*-Tahir to have caught the inverse square law of gravitation if the alien had not done it and in doing so if I had to shake the Mountains of Chagai by hitting my head and take the sky down on earth I would not have refrained from it.

All motion is projectile motion and is the

result of powerful explosions and collisions that happened in the extreme past.

Now let us find the range of the projectile. If we can find the angle θ_0 that the projectile subtends at the centre of the Earth during its flight, then it would become very easy to find the range. So, the main hurdle is the angle θ_0.

Writing the above equation and using Tahir's Principle of Instantaneity, we obtain as follows for *non-gravitivistic* solution of static orbits or projectiles.

$$\frac{d^2r}{dT^2} + \frac{GM}{r^2} = \frac{V^2\cos^2\beta}{r}$$

$$\frac{V^2\cos^2\beta}{r} = r\left[\frac{d\theta}{dT}\right]^2 = \frac{h^2}{r^3}$$

$$\frac{d\theta}{dT} = \frac{h}{r^2}$$

$$\frac{d^2r}{dT^2} + \frac{GM}{r^2} = \frac{h^2}{r^3}$$

$$\frac{d^2r}{dT^2} - \frac{h^2}{r^3} = -\frac{GM}{r^2}$$

$$\frac{dr}{dT} = \frac{dr}{d\theta}\frac{d\theta}{dT} = \frac{h}{r^2}\frac{dr}{d\theta} = h\left[\frac{1}{r^2}\frac{dr}{d\theta}\right]$$

Now we know from chain rule once again that:

$$\frac{d}{d\theta}\left[-\frac{1}{r}\right] = \frac{d}{dr}\left[-\frac{1}{r}\right]\frac{dr}{d\theta} = \frac{1}{r^2}\frac{dr}{d\theta}$$

Therefore

$$\frac{dr}{dT} = h\frac{d}{d\theta}\left[-\frac{1}{r}\right] = -h\frac{d}{d\theta}\left[\frac{1}{r}\right]$$

And

$$\frac{d}{dT}\left[\frac{dr}{dT}\right] = \frac{d^2r}{dT^2} = -h\frac{d}{dT}\frac{d}{d\theta}\left[\frac{1}{r}\right] = -h\frac{d}{d\theta}\frac{d}{d\theta}\left[\frac{1}{r}\right]\frac{d\theta}{dT}$$

$$\frac{d^2r}{dT^2} = -h\frac{d\theta}{dT}\frac{d^2}{d\theta^2}\left[\frac{1}{r}\right] = -\frac{h^2}{r^2}\frac{d^2}{d\theta^2}\left[\frac{1}{r}\right]$$

Now substitute $u = \frac{1}{r}$

$$\frac{d^2r}{dT^2} = -\frac{h^2}{r^2}\frac{d^2u}{d\theta^2}$$

So, we have

$$\frac{d^2r}{dT^2} - \frac{h^2}{r^3} = -\frac{GM}{r^2}$$

$$-\frac{h^2}{r^2}\frac{d^2u}{d\theta^2} - \frac{h^2}{r^3} = -\frac{GM}{r^2}$$

$$\frac{d^2u}{d\theta^2} + u = \frac{GM}{h^2}$$

I will solve this equation differently by utilizing the boundary conditions for the case of a returning projectile which has never been done before. Let us write the equation a bit differently as below.

$$\frac{d^2u}{d\theta^2} + u - \frac{GM}{h^2} = 0$$

Let $f = u - \frac{GM}{h^2}$

$$\frac{df}{d\theta} = \frac{du}{d\theta}$$

$$\frac{d^2f}{d\theta^2} = \frac{d^2u}{d\theta^2}$$

Substituting we obtain

$$\frac{d^2f}{d\theta^2} + f = 0$$

The above equation is a homogeneous second order differential equation. Any reliable website can be checked for their solution.

The characteristic equation for the above differential equation is:

$$w^2 + 1 = 0$$

$$w = \pm i$$

The complementary function is:

$$y_{CF} = C_1\cos\theta + C_2\sin\theta = f$$

We have the final solution as below:

$$u = f + \frac{GM}{h^2}$$

$$u = C_1\cos\theta + C_2\sin\theta + \frac{GM}{h^2}$$

$$\frac{du}{d\theta} = -C_1\sin\theta + C_2\cos\theta$$

$$\frac{d^2u}{d\theta^2} = -C_1\cos\theta - C_2\sin\theta$$

Putting the above two equations back in the alien equation we have

$$-C_1\cos\theta - C_2\sin\theta + C_1\cos\theta + C_2\sin\theta + \frac{GM}{h^2} = \frac{GM}{h^2}$$

$$\frac{GM}{h^2} = \frac{GM}{h^2} \text{ (check)}$$

So, we have

$$u = C_1\cos\theta + C_2\sin\theta + \frac{GM}{h^2}$$

$$\frac{d^2u}{d\theta^2} + u = \frac{GM}{h^2}$$

$$\frac{d^3u}{d\theta^3} + \frac{du}{d\theta} = 0$$

$$\frac{d^4u}{d\theta^4} + \frac{d^2u}{d\theta^2} = 0$$

$$\frac{d^5u}{d\theta^5} + \frac{d^3u}{d\theta^3} = 0$$

.

.

.

.

$$\frac{d^n u}{d\theta^n} + \frac{d^{n-2} u}{d\theta^{n-2}} = 0$$

for n = 3, 4, 5, ... ∞

Now applying the first boundary condition i.e., $\theta = 0$ when $r = R$, we get

$$u = C_1 \cos\theta + C_2 \sin\theta + \frac{GM}{h^2}$$

$$r = \frac{1}{C_1 \cos\theta + C_2 \sin\theta + \frac{GM}{h^2}}$$

$$R = \frac{1}{C_1 + \frac{GM}{h^2}}$$

$$C_1 = \frac{1}{R} - \frac{GM}{h^2}$$

Applying the second boundary condition, i.e., $\theta = \theta_0$ when $r = R$, we get

$$r = \frac{1}{C_1 \cos\theta + C_2 \sin\theta + \frac{GM}{h^2}}$$

$$R = \frac{1}{C_1 \cos\theta_0 + C_2 \sin\theta_0 + \frac{GM}{h^2}}$$

$$R = \frac{1}{(\frac{1}{R} - \frac{GM}{h^2})\cos\theta_0 + C_2 \sin\theta_0 + \frac{GM}{h^2}}$$

$$R\left[(\frac{1}{R} - \frac{GM}{h^2})\cos\theta_0 + C_2 \sin\theta_0 + \frac{GM}{h^2}\right] = 1$$

$$C_2 = \frac{(1 - \cos\theta_0)}{\sin\theta_0}\left(\frac{1}{R} - \frac{GM}{h^2}\right)$$

$$C_2 = \frac{(1 - \cos\theta_0)}{\sin\theta_0} C_1$$

Putting this value of C_2 back in the equation we have

$$r = \frac{1}{C_1 \cos\theta + \frac{(1 - \cos\theta_0)}{\sin\theta_0} C_1 \sin\theta + \frac{GM}{h^2}}$$

Simplifying we get

$$r = \frac{\sin\theta_0}{C_1[\sin(\theta_0 - \theta) + \sin\theta] + \frac{GM}{h^2}\sin\theta_0}$$

Before proceeding any further, let us first find the maximum radial distance that the projectile will travel before returning towards the Earth or the gravitating mass as below by using the Tahirian equations.

$$h = rv\cos\beta$$

$$\frac{V_2^2}{2} - \frac{V_1^2}{2} = GM(\frac{1}{r_2} - \frac{1}{r_1})$$

$$h = r_0 V_0 \cos\beta_0 = R v_R' \cos\beta_R = r_0 V_0$$

$$V_0 = \frac{h}{r_0}$$

Where $\beta_0 = 0$ at the turning point of the projectile.

Inserting this value of v_0 in the above Tahir's Work-Energy theorem, which has been under suspicious commentary by aliens in the corners of the World, we obtain:

$$\frac{V_2^2}{2} - \frac{V_1^2}{2} = GM(\frac{1}{r_2} - \frac{1}{r_1})$$

$$V_2^2 - V_1^2 = 2GM(\frac{1}{r_2} - \frac{1}{r_1})$$

$$\frac{h^2}{r_0^2} - V_R^2 = 2GM(\frac{1}{r_0} - \frac{1}{R})$$

We finally obtain:

$$A r_0^2 + 2B r_0 - h^2 = 0$$

The above equation seems to be a mere equation, it is the destiny of the sons and daughters of Adam and Hawwa in reality!

Where $A = V_R^2 - 2\frac{GM}{R}$, $B = GM$ & $h = R V_R^2 \cos\beta_R$
We will also see next that the gravitivistic counterpart is as below:

$$v_{2r_0}^2 - v_{2R}^2 = -\frac{2Gm_1}{t_2^2 r_{21}} + \frac{2Gm_1}{t_2^2 r_{2r_0}} + \frac{6G^2 m_1^2}{t_2^2 s^2 r_{2r_0}^2} - \frac{12G^2 m_1^2}{t_2^2 s^2 r_{21} r_{2r_0}}$$
$$+ \frac{6G^2 m_1^2}{t_2^2 s^2 r_{21}^2} + \frac{6Gm_1 v_{21}^2}{s^2 r_{2r_0}} - \frac{6Gm_1 v_{21}^2}{s^2 r_{21}} + \cdots$$

Now we know that:

$$V_{2r_0} = \frac{h_2}{r_{2r_0}}$$

$$\frac{h_2^2}{r_{2r_0}^2} - v_{2R}^2 = -\frac{2Gm_1}{t_2^2 r_{21}} + \frac{2Gm_1}{t_2^2 r_{2r_0}} + \frac{6G^2 m_1^2}{t_2^2 s^2 r_{2r_0}^2} - \frac{12G^2 m_1^2}{t_2^2 s^2 r_{21} r_{2r_0}}$$
$$+ \frac{6G^2 m_1^2}{t_2^2 s^2 r_{21}^2} + \frac{6Gm_1 v_{21}^2}{s^2 r_{2r_0}} - \frac{6Gm_1 v_{21}^2}{s^2 r_{21}} + \cdots$$

$$h^2 \approx (v_{2R}^2 - \frac{2Gm_1}{t_2^2 r_{21}} - \frac{6Gm_1 v_{21}^2}{s^2 r_{21}} + \frac{6G^2 m_1^2}{t_2^2 s^2 r_{21}^2})^2 r_{2r_0}$$

$$+\left(\frac{2Gm_1}{t_2^2} - \frac{12G^2m_1^2}{t_2^2 s^2 r_{21}} + \frac{6Gm_1 v_{21}^2}{s^2}\right)r_{2r_0} + \frac{6G^2m_1^2}{t_2^2 s^2}$$

$$\left(v_{2R}^2 - \frac{2Gm_1}{t_2^2 r_{21}} - \frac{6Gm_1 v_{21}^2}{s^2 r_{21}} + \frac{6G^2m_1^2}{t_2^2 s^2 r_{21}^2}\right)r_{2r_0}^2$$
$$+ \left(\frac{2Gm_1}{t_2^2} - \frac{12G^2m_1^2}{t_2^2 s^2 r_{21}} + \frac{6Gm_1 v_{21}^2}{s^2}\right)r_{2r_0}$$
$$+ \frac{6G^2m_1^2}{t_2^2 s^2} - h_2^2 \approx 0$$

Where:

$$A = v_{2R}^2 - \frac{2Gm_1}{t_2^2 r_{2R}} - \frac{6Gm_1 v_{21}^2}{s^2 r_{2R}} + \frac{6G^2m_1^2}{t_2^2 s^2 r_{2R}^2}$$

$$B = \frac{Gm_1}{t_2^2} - \frac{6G^2m_1^2}{t_2^2 s^2 r_{2R}} + \frac{3Gm_1 v_{21}^2}{s^2}$$

$$C = \frac{6G^2m_1^2}{t_2^2 s^2} - h_2^2$$

$$h = RV_R \cos\beta_R$$
$$r_{2r_0} = r_{22}$$
$$r_{2R} = r_{21}$$

Care should be exercised because I have used a, b & c to represent different expressions at different places interchangeably. I thought they would be different, but they are the same. Solving the above non-gravitivistic and gravitivistic equation by Al-Khawarizmi's Quadratic formula we obtain with negative sign as below:

$$r_0 = \frac{B + \sqrt{B^2 + AC^2}}{-A}$$

This is the maximum radial distance that the projectile will travel from the center of the Earth before returning to the Earth or the gravitating mass at all Tahirian Speeds lower than escape speed.

$$r = \frac{\sin\theta_0}{\left[\frac{1}{R} - \frac{GM}{h^2}\right][\sin(\theta_0 - \theta) + \sin\theta] + \frac{GM}{h^2}\sin\theta_0}$$

If we differentiate r in the above equation with respect to either t or θ and put, $\frac{dr}{dt}$ or $\frac{dr}{d\theta} = 0$, we obtain $\theta = \frac{\theta_0}{2}$. Only $\frac{d\theta}{dt} \neq 0$ at $\theta = \frac{\theta_0}{2}$. R_0 occurs at $\theta = \frac{\theta_0}{2}$.

So, putting the value of $\theta = \frac{\theta_0}{2}$ back into the equation and solving for θ_0, we obtain,

$$\theta_0 = 2\cos^{-1}\left[r_0 \frac{\left(\frac{1}{R} - \frac{GM}{h^2}\right)}{\left(1 - r_0\frac{GM}{h^2}\right)}\right]$$

So, the equation of the projectile is:

$$r = \frac{\sin\theta_0}{\left[\frac{1}{R} - \frac{GM}{h^2}\right][\sin(\theta_0 - \theta) + \sin\theta] + \frac{GM}{h^2}\sin\theta_0}$$

Where

$$\theta_0 = 2\cos^{-1}\left[r_0 \frac{\left(\frac{1}{R} - \frac{GM}{h^2}\right)}{\left(1 - r_0\frac{GM}{h^2}\right)}\right]$$

For non-gravitivistic value of r_0, we obtain

$$r_0 = \frac{B + \sqrt{B^2 + Ah^2}}{-A}$$

$$r_0 = \frac{GM + \left\{G^2M^2 + [v_R'^2 - 2\frac{GM}{R}]R^2 v_R^2 \cos^2\beta_R\right\}^{1/2}}{2\frac{GM}{R} - v_R^2}$$

And for the gravitivistic value we can use the gravitivistic values from A, B & C. I haven't used the Tahir Constant of Binary Stars cosmology i.e., t in the non-gravitivistic values and have kept it as part of the CRUX of Gravitivity because I believe Gravitivity is one-fifth of Tahirian Gravitation. The only difference is that we will substitute the constant h by C as below:

$$r_0 = \frac{B + \sqrt{B^2 + AC^2}}{-A}$$

The non-gravitivistic range of the projectile, for perfectly spherical earth, can be found by the formula as below:

$$S = R\theta_0$$

Where, θ_0 is the angle subtended by the projectile at the Centre of the Earth. Hence

$$S = R\theta_0 = 2R\cos^{-1}\left[r_0 \frac{\left(\frac{1}{R} - \frac{GM}{h^2}\right)}{\left(1 - r_0\frac{GM}{h^2}\right)}\right]$$

Now, in order, to find the optimal range at a certain speed, it is very important to consider everything in terms of r_0 & v_0. So, writing the Tahirian Constraint Equation of Inertia and Gravitation as below we have

$$Ar_0^2 + 2Br_0 - h^2 = 0$$
$$Ar_0^2 + 2Br_0 - r_0^2 v_0^2 = 0$$

Or

$$Ar_0 + 2B - r_0 v_0^2 = 0$$

Or

$$g(r_0, v_0) = Ar_0 + 2B - r_0 v_0^2 = 0$$

Tahirian Constraint Equation of Inertia and Gravitation

And the Principal quantity to be optimized is as below:

$$S = 2R\cos^{-1}[r_0 \frac{\left(\frac{1}{R} - \frac{GM}{h^2}\right)}{\left(1 - r_0\frac{GM}{h^2}\right)}]$$

Where:

$$h = r_0 v_0 \cos\beta_0 = Rv_R\cos\beta_R = r_0 v_0 = h(r_0, v_0)$$

We CAN'T consider h a constant during the process of partial differentiation because we must confer full freedom to h to change and hence to the product $r_0 v_0$. We are optimizing the angle of launch of the projectile at which it reaches the maximum distance/range and when the angle is considered changing, we can't consider h as constant prior to differentiating! So, we have the equation for Tahirian Cosmology as below:

$$S(r_0, v_0) = 2R\cos^{-1}[\frac{\left(\frac{r_0}{R} - \frac{B}{r_0 v_0^2}\right)}{\left(1 - \frac{B}{r_0 v_0^2}\right)}]$$

$$g(r_0, v_0) = Ar_0 + 2B - r_0 v_0^2 = 0$$

At the maximum value of the function $S(r_0, v_0)$ with the constraint equation $g(r_0, v_0)$, we have the gradient vectors parallel to each other. When two vectors are parallel, we can connect them with the help of a constant. Let us call the constant λ. Therefore, we obtain:

$$\nabla S(r_0, v_0) = \lambda \nabla g(r_0, v_0)$$

The above equation can be further divided into two equations as below:

$$\frac{\partial}{\partial r_0} S(r_0, v_0) = \lambda \frac{\partial}{\partial r_0} g(r_0, v_0)$$

&

$$\frac{\partial}{\partial v_0} S(r_0, v_0) = \lambda \frac{\partial}{\partial v_0} g(r_0, v_0)$$

Substituting value of λ from one equation into the other we have

$$\frac{\frac{\partial}{\partial r_0} S(r_0, v_0)}{\frac{\partial}{\partial v_0} S(r_0, v_0)} = \frac{\frac{\partial}{\partial r_0} g(r_0, v_0)}{\frac{\partial}{\partial v_0} g(r_0, v_0)}$$

So, we have

$$\frac{\partial}{\partial r_0} S(r_0, v_0) = \frac{-2R}{\sqrt{1 - (\frac{\frac{r_0}{R} - \frac{B}{r_0 v_0^2}}{1 - \frac{B}{r_0 v_0^2}})^2}} \frac{(\frac{1}{R} + \frac{B}{h^2} - 2\frac{Br_0}{Rh^2})}{(1 - \frac{B}{r_0 v_0^2})^2}$$

&

$$\frac{\partial}{\partial v_0} S(r_0, v_0) = \frac{-2R}{\sqrt{1 - (\frac{\frac{r_0}{R} - \frac{B}{r_0 v_0^2}}{1 - \frac{B}{r_0 v_0^2}})^2}} \frac{(1 - \frac{r_0}{R})2\frac{B}{r_0 v_0^3}}{(1 - \frac{B}{r_0 v_0^2})^2}$$

$$\frac{\partial}{\partial r_0} g(r_0, v_0) = A - v_0^2$$

&

$$\frac{\partial}{\partial v_0} g(r_0, v_0) = -2h$$

Putting the partial derivatives in the equation before we have

$$\frac{\frac{1}{R} + \frac{B}{h^2} - 2\frac{Br_0}{Rh^2}}{(1 - \frac{r_0}{R})2\frac{B}{r_0 v_0^3}} = \frac{A - v_0^2}{-2h}$$

$$\frac{(\frac{1}{R} + \frac{B}{h^2} - 2\frac{Br_0}{Rh^2})r_0 v_0^3}{(1 - \frac{r_0}{R})2B} = \frac{A - v_0^2}{-2h}$$

Multiplying both sides by r_0^2, we obtain

$$\frac{(\frac{1}{R} + \frac{B}{h^2} - 2\frac{Br_0}{Rh^2})r_0^3 v_0^3}{(1 - \frac{r_0}{R})2B} = \frac{Ar_0^2 - r_0^2 v_0^2}{-2h}$$

$$\frac{(\frac{1}{R} + \frac{B}{h^2} - 2\frac{Br_0}{Rh^2})h^3}{(1 - \frac{r_0}{R})2B} = \frac{Ar_0^2 - h^2}{-2h}$$

Now we know from before that:

$$Ar_0^2 + 2Br_0 - h^2 = 0$$

Or

$$Ar_0^2 - h^2 = -2Br_0$$

So, we have

$$\frac{(\frac{1}{R} + \frac{B}{h^2} - 2\frac{Br_0}{Rh^2})h^3}{(1 - \frac{r_0}{R})2B} = \frac{-2Br_0}{-2h} = \frac{Br_0}{h}$$

$$\left(\frac{1}{R} + \frac{B}{h^2} - 2\frac{Br_0}{Rh^2}\right)h^4 = 2B^2r_0(1 - \frac{r_0}{R})$$

Now let $p = h^2$

$$\left(\frac{1}{R} + \frac{B}{p} - 2\frac{Br_0}{Rp}\right)p^2 = 2B^2r_0(1 - \frac{r_0}{R})$$

$$\left(1 + \frac{BR}{p} - 2\frac{Br_0}{p}\right)p^2 = 2B^2r_0(R - r_0)$$

$$p^2 + BRp - 2Br_0p = 2B^2r_0R - 2B^2r_0^2$$

Rearranging terms

$$2B^2r_0R - 2B^2r_0^2 = p^2 + BRp - 2Br_0p$$

Multiplying by -1

$$2B^2r_0^2 - 2B^2r_0R = 2Br_0p - BRP - p^2$$

Multiplying by $\frac{A}{2B^2}$

$$Ar_0^2 - ARr_0 = \frac{A}{B}pr_0 - \frac{A}{2B}Rp - \frac{A}{2B^2}p^2$$

Rearranging terms

$$Ar_0^2 + \left(-AR - \frac{A}{B}p\right)r_0 = -(\frac{A}{2B}Rp + \frac{A}{2B^2}p^2)$$

Comparing with and equating coefficients of like powers of r_0, we have as below:

$$Ar_0^2 + \left(-AR - \frac{A}{B}p\right)r_0 = -(\frac{A}{2B}Rp + \frac{A}{2B^2}p^2)$$

$$Ar_0^2 + 2Br_0 = h^2 = p$$

$$-AR - \frac{A}{B}p = 2B$$

&

$$-\left(\frac{A}{2B}Rp + \frac{A}{2B^2}p^2\right) = p$$

Solving any of the two equations just above for p produces the same value of p as below:

$$p = \frac{(2B + AR)}{-A}B$$

Now we know that:

$$h^2 = p$$

$$h^2 = \frac{(2B + AR)}{-A}B$$

$$h = \sqrt{\frac{(2B + AR)}{-A}B}$$

$$h = Rv_R\cos\beta_R = \sqrt{\frac{(2B + AR)}{-A}B}$$

$$\cos\beta_R = \frac{1}{Rv_R}\sqrt{\frac{(2B + AR)}{-A}B}$$

$$\cos\beta_R = \frac{1}{Rv_R}\sqrt{\frac{(2B + AR)}{-A}B}$$

For those who like to have non-gravitivistic value, they can put non-gravitivistic values of A and B. But for those who want to comfort their eyes, for the first time in the Tahirian Era, to see the Tahirian value, they can solve the differential equation! For non-gravitivistic kids we have

$$\cos\beta_R = \frac{1}{Rv_R}\sqrt{[-\frac{2GM}{v_R^2 - \frac{2GM}{R}} - R]GM}$$

$$\cos\beta_R = \frac{1}{Rv_R}\sqrt{[\frac{2GM}{\frac{2GM}{R} - v_R^2} - R]GM}$$

$$\cos\beta_R = \frac{1}{Rv_R}\sqrt{\left[\frac{1}{1 - \frac{v_R^2R}{2GM}} - 1\right]GMR}$$

$$\cos\beta_R = \sqrt{\frac{GM}{Rv_R^2}\left[\frac{1}{1 - \frac{v_R^2R}{2GM}} - 1\right]}$$

Care should be exercised in that the denominator term is approximated by using Umer Khayyam's binomial series expansion before the two terms are subtracted together for showing $\beta_R \approx 45$ degrees at low speed.

This is the optimal value (non-gravitivistic) of the angle of launch of a projectile at all Tahirian speeds tangential to the surface of the Earth or of any spherical gravitating matter for maximum range.

For $0 < v <<< \sqrt{\frac{2GM}{R}} \Rightarrow \beta_R \approx 45$ degrees.

Natural Theory of Relativity, Inertia, Gravitation & Gravitivity

For $v = \sqrt{\frac{GM}{R}} \Rightarrow \beta_R = 0$ degrees.

For velocities in the range as below

$$\sqrt{\frac{GM}{R}} \le v < \sqrt{\frac{2GM}{R}}$$

β_R stays equal to zero i.e., $\beta_R = 0$ for maximum range! The only difference is that the orbits are non-circular now. This is also popular among aliens as a source of recreational balls. In the *approximately* elliptical, parabolic and hyperbolic scenarios the body doesn't fall back on earth because of which the scenarios are not covered by the Tahirian equation.

θ is taken +ve in the counterclockwise direction as per one of the rules of Alkashi's Trigonometry

For the non-gravitivistic Tahirian Value of the optimal angle of launch of the projectile that produces the maximum range of the projectile, we must use the Tahir-alien equation as below:

$$\frac{d^2u}{d\theta^2} + u = \frac{GM}{h^2} \text{ (alien equation)}$$

$$\frac{d^2u}{d\theta^2} + u = (1 + 3t_2^2 \frac{v_{22}^2}{s^2}) \frac{GM}{h^2} \text{ (Tahir-Alien Equation)}$$

$$1 + 3t_2^2 \frac{v_{22}^2}{s^2} = \left(1 + 3t_2^2 \frac{v_{21}^2}{s^2}\right) e^{[\frac{6Gm_1}{s^2}(\frac{1}{r_{22}} - \frac{1}{r_{21}})]}$$

Now using the Tahirian substitution as below

$$t_2 = 1 + \frac{m_2}{m_1}$$

$$r^2 = r_2^2 (1 + \frac{m_2}{m_1})^2 = r_2^2 t_2^2$$

We modify the above Tahir-Alien Equation as per the Canons of Tahirian Cosmology as below:

$$\frac{d^2r}{dT^2} - \frac{h^2}{r^3} = -\frac{GM}{r^2} \text{ (ae)}$$

$$\frac{d^2r_{22}}{dT^2} - \frac{h_2^2}{r_{22}^3} = -(1 + 3t_2^2 \frac{v_{22}^2}{s^2}) \frac{GM}{r^2} \text{ (TAE)}$$

$$-\frac{h_2^2}{r_{22}^2} \frac{d^2u_{22}}{d\theta^2} - \frac{h_2^2}{r_{22}^3} = (1 + 3t_2^2 \frac{v_{22}^2}{s^2}) \frac{GM}{r^2}$$

$$\frac{d^2u_{22}}{d\theta^2} + u_{22} = (1 + 3t_2^2 \frac{v_{22}^2}{s^2}) \frac{Gm_1}{t_2^2 h_2^2}$$

Where:

$$u_{22} = \frac{1}{r_{22}}$$

Substituting the value of $(1 + 3t_2^2 \frac{v_{22}^2}{s^2}) \frac{GM}{r^2}$, we obtain as below:

$$\frac{d^2u_{22}}{d\theta^2} + u_{22} = \left(1 + 3t_2^2 \frac{v_{21}^2}{s^2}\right) e^{[\frac{6Gm_1}{s^2}(\frac{1}{r_{22}} - \frac{1}{r_{21}})]} \frac{Gm_1}{t_2^2 h_2^2}$$

Expanding the exponential function as per Tahir Series Expansion of functions, we obtain

$$\frac{d^2u_{22}}{d\theta^2} + u_{22} = \left(1 + 3t_2^2 \frac{v_{21}^2}{s^2}\right)(1 + \frac{6Gm_1}{s^2}(u_{22} - u_{21}) \frac{Gm_1}{t_2^2 h_2^2}$$

Calculations with powers of s higher than 2 in the denominator are left as an exercise for the aliens.

$$\frac{d^2u_{22}}{d\theta^2} + u_{22} = \left(1 + 3t_2^2 \frac{v_{21}^2}{s^2}\right)(1 + \frac{6Gm_1}{s^2}u_{22} - \frac{6Gm_1}{s^2}u_{21}) \frac{Gm_1}{t_2^2 h_2^2}$$

$$\frac{d^2u_{22}}{d\theta^2} + u_{22} = (1 + \frac{6Gm_1}{s^2}u_{22} - \frac{6Gm_1}{s^2}u_{21} + 3t_2^2 \frac{v_{21}^2}{s^2}) \frac{Gm_1}{t_2^2 h_2^2}$$

Rearranging like terms, we obtain

$$\frac{d^2u_{22}}{d\theta^2} + u_{22}(1 - \frac{6G^2m_1^2}{t_2^2 h_2^2 s^2}) = (1 + 3t_2^2 \frac{v_{21}^2}{s^2} - \frac{6Gm_1}{s^2}u_{21}) \frac{Gm_1}{t_2^2 h_2^2}$$

Let $p = 1 - \frac{6G^2m_1^2}{t_2^2 h_2^2 s^2}$ & $q = 1 + 3t_2^2 \frac{v_{21}^2}{s^2} - \frac{6Gm_1}{s^2}u_{21}$

$$\frac{d^2u_{22}}{d\theta^2} + pu_{22} = q\frac{Gm_1}{t_2^2 h_2^2}$$

Let us write the equation as below:

$$\frac{d^2u_{22}}{d\theta^2} + pu_{22} - q\frac{Gm_1}{t_2^2 h_2^2} = 0$$

Let $f_{22} = pu_{22} - q\frac{Gm_1}{t_2^2 h_2^2}$

$$\frac{df_{22}}{d\theta} = p\frac{du_{22}}{d\theta}$$

$$\frac{d^2f_{22}}{d\theta^2} = p\frac{d^2u_{22}}{d\theta^2}$$

$$\frac{1}{p}\frac{d^2f_{22}}{d\theta^2} = \frac{d^2u_{22}}{d\theta^2}$$

Substituting we obtain

$$\frac{1}{p}\frac{d^2f_{22}}{d\theta^2} + f = 0$$

$$\frac{d^2f_{22}}{d\theta^2} + pf = 0$$

The above equation is a homogeneous second order differential equation. Any reliable website can be checked for their solution.

The characteristic equation for the above differential equation is:

$$w^2 + p = 0$$

$$w = \pm i\sqrt{p}$$

The complementary function is:

$$y_{CF} = C_1 \cos\sqrt{p}\theta + C_2 \sin\sqrt{p}\theta = f_{22}$$

We have the final solution, as below:

$$u_{22} = \frac{C_1}{p}\cos\sqrt{p}\theta + \frac{C_2}{p}\sin\sqrt{p}\theta + \frac{q}{p}\frac{GM}{t_2^2 h^2}$$

$$u_{22} = C_1'\cos\sqrt{p}\theta + C_2'\sin\sqrt{p}\theta + \frac{q}{p}\frac{GM}{t_2^2 h^2}$$

Where $C_1' = \frac{C_1}{p}$ & $C_2' = \frac{C_2}{p}$

$$\frac{du_{22}}{d\theta} = \{-C_1'\sin\sqrt{p}\theta + C_2'\cos\sqrt{p}\theta\}\sqrt{p}$$

$$\frac{d^2 u_{22}}{d\theta^2} = \{-C_1'\cos\sqrt{p}\theta - C_2'\sin\sqrt{p}\theta\}p$$

Putting the above two equations back in the alien equation we have

$$\{-C_1'\cos\sqrt{p}\theta - C_2'\sin\sqrt{p}\theta\}p + p(C_1'\cos\sqrt{p}\theta + C_2'\sin\sqrt{p}\theta + \frac{q}{p}\frac{GM}{t_2^2 h^2}) = q\frac{GM}{t_2^2 h^2}$$

$$q\frac{GM}{t_2^2 h^2} = q\frac{GM}{t_2^2 h^2} \quad \textbf{(check)}$$

Now applying the first boundary condition i.e., $\theta = 0$ when $r_{22} = R$, we get

$$u_{22} = C_1'\cos\sqrt{p}\theta + C_2'\sin\sqrt{p}\theta + \frac{q}{p}\frac{GM}{t_2^2 h^2}$$

$$r_{22} = \frac{1}{C_1'\cos\sqrt{p}\theta + C_2'\sin\sqrt{p}\theta + \frac{q}{p}\frac{GM}{t_2^2 h^2}}$$

$$R = \frac{1}{C_1' + \frac{q}{p}\frac{GM}{t_2^2 h^2}}$$

$$C_1' = \frac{1}{R} - \frac{q}{p}\frac{GM}{t_2^2 h^2}$$

Here we notice that the term:

$$q\frac{Gm_1}{t_2^2} = (1 + 3t_2^2\frac{v_{21}^2}{s^2} - \frac{6Gm_1}{s^2}u_{21})\frac{Gm_1}{t_2^2} = B$$

$$B = \frac{Gm_1}{t_2^2} - \frac{6G^2 m_1^2}{t_2^2 s^2 r_{2R}} + \frac{3Gm_1 v_{21}^2}{s^2} \textbf{ (Gravitivistic term; Binary term)}$$

$$B = GM \textbf{ (non − Gravitivistic; non − Binary; alien)}$$

Applying the second boundary condition, i.e., $\theta = \theta_0$ when $r_{22} = R$, we get

$$r_{22} = \frac{1}{C_1'\cos\sqrt{p}\theta + C_2'\sin\sqrt{p}\theta + \frac{B}{ph^2}}$$

$$C_1' = \frac{1}{R} - \frac{B}{ph^2}$$

$$R = \frac{1}{C_1'\cos\sqrt{p}\theta_0 + C_2'\sin\sqrt{p}\theta_0 + \frac{B}{ph^2}}$$

$$R = \frac{1}{(\frac{1}{R} - \frac{B}{ph^2})\cos\sqrt{p}\theta_0 + C_2'\sin\sqrt{p}\theta_0 + \frac{B}{ph^2}}$$

$$R\left[(\frac{1}{R} - \frac{B}{ph^2})\cos\sqrt{p}\theta_0 + C_2'\sin\sqrt{p}\theta_0 + \frac{B}{ph^2}\right] = 1$$

$$C_2' = \frac{(1 - \cos\sqrt{p}\theta_0)}{\sin\sqrt{p}\theta_0}\left(\frac{1}{R} - \frac{B}{ph^2}\right)$$

$$C_2' = \frac{(1 - \cos\sqrt{p}\theta_0)}{\sin\sqrt{p}\theta_0}C_1'$$

Putting this value of C_2' back in the equation we have

$$r = \frac{1}{C_1'\cos\sqrt{p}\theta + \frac{(1 - \cos\sqrt{p}\theta_0)}{\sin\sqrt{p}\theta_0}C_1'\sin\sqrt{p}\theta + \frac{B}{ph^2}}$$

Simplifying we get

$$r_{22} = \frac{\sin\sqrt{p}\theta_0}{C_1'[\sin\sqrt{p}(\theta_0 - \theta) + \sin\sqrt{p}\theta] + \frac{B}{ph^2}\sin\sqrt{p}\theta_0}$$

This is Tahir-Alien equation of the trajectories of Tahirian projectiles and planets. Where are the copernicoons and their revolutions, where are the golileo polios and their naïve thoughts? Another good name for my theory can be "**Theory of Binary Stars**".

If we differentiate r_{22} in the above equation with respect to either t or θ and put, $\frac{dr}{dt}$ or $\frac{dr}{d\theta} = 0$, we obtain $\theta = \frac{\theta_0}{2}$. Only $\frac{d\theta}{dt} \neq 0$ at $\theta = \frac{\theta_0}{2}$. r_0 occurs at $\theta = \frac{\theta_0}{2}$.

So, putting the value of $\theta = \frac{\theta_0}{2}$ back into the equation and solving for θ_0, we obtain,

$$\theta_0 = \frac{2}{\sqrt{p}}\cos^{-1}[r_0\frac{\left(\frac{1}{R} - \frac{B}{ph^2}\right)}{\left(1 - r_0\frac{B}{ph^2}\right)}]$$

The gravitivistic range of the projectile, for perfectly spherical earth, can be found by the formula as below:

$$S = R\theta_0$$

Where, θ_0 is the angle subtended by the projectile at the Centre of the Earth. Hence

$$S = R\theta_0 = \frac{2}{\sqrt{p}} R\cos^{-1}[r_0 \frac{\left(\frac{1}{R} - \frac{B}{ph^2}\right)}{\left(1 - r_0 \frac{B}{ph^2}\right)}]$$

Now, in order, to find the optimal range at a certain speed, it is very important to consider everything in terms of r_0 & v_0. So, writing the Tahirian Constraint Equation *of* Inertia and Gravitation as below we have

$$A = v_{2R}^2 - \frac{2Gm_1}{t_2^2 r_{2R}} - \frac{6Gm_1 v_{21}^2}{s^2 r_{2R}} + \frac{6G^2 m_1^2}{t_2^2 s^2 r_{2R}^2}$$

$$B = \frac{Gm_1}{t_2^2} - \frac{6G^2 m_1^2}{t_2^2 s^2 r_{2R}} + \frac{3Gm_1 v_{21}^2}{s^2}$$

$$C = \frac{6G^2 m_1^2}{t_2^2 s^2} - h_2^2 = -h_2^2\left(1 - \frac{6G^2 m_1^2}{t_2^2 s^2 h_2^2}\right) = -h_2^2 p$$

$$p = 1 - \frac{6G^2 m_1^2}{t_2^2 s^2 h_2^2}$$

$$h_2 = r_0 v_0$$
$$r_{2r_0} = r_{22} = r_0$$
$$r_{2R} = r_{21} = R$$

$$Ar_0^2 + 2Br_0 - h_2^2 p = 0$$

$$Ar_0^2 + 2Br_0 - r_0^2 v_0^2 p = 0$$

Or

$$Ar_0 + 2B - r_0 v_0^2 p = 0$$

Or

$$g(r_0, v_0) = Ar_0 + 2B - r_0 v_0^2 p = 0$$

(Tahirian Constraint Equation *of* Inertia and Gravitation)

The principal quantity to be optimized is as below:

$$S = R\theta_0 = \frac{2}{\sqrt{p}} R\cos^{-1}[r_0 \frac{\left(\frac{1}{R} - \frac{B}{ph_2^2}\right)}{\left(1 - r_0 \frac{B}{ph_2^2}\right)}]$$

Where
$$h_2 = r_0 v_0 \cos\beta_0 = r_0 v_0$$

We CAN'T consider h a constant during the process of partial differentiation because we have to confer full freedom to h in order to change and hence to the product $r_0 v_0$. We are optimizing the angle of launch of the projectile at which it reaches the maximum distance/range and when the angle is considered changing, we can't consider h as constant prior to differentiating! So, we have the equations for Tahirian Cosmology as below:

$$S(r_0, v_0) = 2R\cos^{-1}[\frac{\left(\frac{r_0}{R} - \frac{B}{r_0 v_0^2 p}\right)}{\left(1 - \frac{B}{r_0 v_0^2 p}\right)}]$$

$$g(r_0, v_0) = Ar_0 + 2B - r_0 v_0^2 p = 0$$

At the maximum value of the function $S(r_0, v_0)$ with the constraint equation $g(r_0, v_0)$, we have the gradient vectors parallel to each other. When two vectors are parallel, we can connect them with the help of a constant. Let us call the constant λ. Therefore, we obtain as before.

$$\nabla S(r_0, v_0) = \lambda \nabla g(r_0, v_0)$$

The above equation can be further divided into two equations as below:

$$\frac{\partial}{\partial r_0} S(r_0, v_0) = \lambda \frac{\partial}{\partial r_0} g(r_0, v_0)$$

&

$$\frac{\partial}{\partial v_0} S(r_0, v_0) = \lambda \frac{\partial}{\partial v_0} g(r_0, v_0)$$

Substituting value of λ from one equation into the other we have

$$\frac{\frac{\partial}{\partial r_0} S(r_0, v_0)}{\frac{\partial}{\partial v_0} S(r_0, v_0)} = \frac{\frac{\partial}{\partial r_0} g(r_0, v_0)}{\frac{\partial}{\partial v_0} g(r_0, v_0)}$$

So, we have

$$\frac{\partial}{\partial r_0} S(r_0, v_0)$$

$$= \frac{-2R}{\sqrt{1 - (\frac{\left(\frac{r_0}{R} - \frac{B}{r_0 v_0^2 p}\right)}{\left(1 - \frac{B}{r_0 v_0^2 p}\right)})^2}} \frac{\left(\frac{1}{R} + \frac{B}{ph^2} - 2\frac{Br_0}{pRh^2}\right)}{(1 - \frac{B}{r_0 v_0^2 p})^2}$$

&

$$\frac{\partial}{\partial v_0} S(r_0, v_0) = \frac{-2R}{\sqrt{1 - \left(\frac{\frac{r_0}{R} - \frac{B}{pr_0 v_0^2}}{1 - \frac{B}{pr_0 v_0^2}}\right)^2}} \cdot \frac{(1 - \frac{r_0}{R})2\frac{B}{pr_0 v_0^3}}{(1 - \frac{B}{pr_0 v_0^2})^2}$$

$$\frac{\partial}{\partial r_0} g(r_0, v_0) = A - v_0^2 p$$

&

$$\frac{\partial}{\partial v_0} g(r_0, v_0) = -2h_2 p$$

Putting the partial derivatives in the equation before we have

$$\frac{\frac{1}{R} + \frac{B}{ph^2} - 2\frac{Br_0}{pRh^2}}{(1 - \frac{r_0}{R})2\frac{B}{pr_0 v_0^3}} = \frac{A - v_0^2 p}{-2hp}$$

$$\frac{(\frac{1}{R} + \frac{B}{ph^2} - 2\frac{Br_0}{pRh^2})r_0 v_0^3 p}{(1 - \frac{r_0}{R})2B} = \frac{A - v_0^2 p}{-2hp}$$

Multiplying both sides by r_0^2, we obtain

$$\frac{(\frac{1}{R} + \frac{B}{ph^2} - 2\frac{Br_0}{pRh^2})r_0^3 v_0^3 p}{(1 - \frac{r_0}{R})2B} = \frac{Ar_0^2 - r_0^2 v_0^2 p}{-2hp}$$

$$\frac{(\frac{1}{R} + \frac{B}{ph^2} - 2\frac{Br_0}{pRh^2})h^3 p}{(1 - \frac{r_0}{R})2B} = \frac{Ar_0^2 - h^2 p}{-2hp}$$

Now we know from before that:

$$Ar_0^2 + 2Br_0 - h^2 p = 0$$

Or

$$Ar_0^2 - h^2 p = -2Br_0$$

So, we have

$$\frac{(\frac{1}{R} + \frac{B}{ph^2} - 2\frac{Br_0}{pRh^2})h^3 p}{(1 - \frac{r_0}{R})2B} = \frac{-2Br_0}{-2hp} = \frac{Br_0}{hp}$$

$$\left(\frac{1}{R} + \frac{B}{ph^2} - 2\frac{Br_0}{pRh^2}\right)h^4 p = \frac{2B^2 r_0}{p}(1 - \frac{r_0}{R})$$

Now let $H = h^2$

$$\left(\frac{1}{R} + \frac{B}{pH} - 2\frac{Br_0}{pRH}\right)H^2 p = \frac{2B^2 r_0}{p}(1 - \frac{r_0}{R})$$

$$\left(1 + \frac{BR}{pH} - 2\frac{Br_0}{pH}\right)H^2 p^2 = 2B^2 r_0(R - r_0)$$

$$p^2 H^2 + pBRH - 2pBr_0 H = 2B^2 r_0 R - 2B^2 r_0^2$$

Rearranging terms

$$2B^2 r_0 R - 2B^2 r_0^2 = p^2 H^2 + pBRH - 2pBr_0 H$$

Multiplying by -1

$$2B^2 r_0^2 - 2B^2 r_0 R = 2pBr_0 H - pBRH - p^2 H^2$$

Multiplying by $\frac{A}{2B^2}$

$$Ar_0^2 - ARr_0 = \frac{A}{B}pr_0 H - \frac{A}{2B}RpH - \frac{A}{2B^2}p^2 H^2$$

Rearranging terms

$$Ar_0^2 + \left(-AR - \frac{A}{B}pH\right)r_0 = -\left(\frac{A}{2B}RpH + \frac{A}{2B^2}p^2 H^2\right)$$

Comparing and equating coefficients of like powers of r_0, we have as below:

$$Ar_0^2 + \left(-AR - \frac{A}{B}pH\right)r_0 = -\left(\frac{A}{2B}RpH + \frac{A}{2B^2}p^2 H^2\right)$$

$$Ar_0^2 + 2Br_0 = pH$$

$$-AR - \frac{A}{B}pH = 2B$$

&

$$-\left(\frac{A}{2B}RpH + \frac{A}{2B^2}p^2 H^2\right) = pH$$

Solving any of the two equations just above for p produces the same value of p as below:

$$H = \frac{(2B + AR)}{-Ap}B$$

Now, we know that:

$$h^2 = H$$

$$h^2p = \frac{(2B + AR)}{-A}B$$

$$h_2^2 - \frac{6G^2m_1^2}{t_2^2s^2} = \frac{(2B + AR)}{-A}B$$

$$h_2^2 = \frac{(2B + AR)}{-A}B + \frac{6G^2m_1^2}{t_2^2s^2}$$

$$h_2 = \sqrt{\frac{(2B + AR)}{-A}B + \frac{6G^2m_1^2}{t_2^2s^2}}$$

Let $d = \frac{6G^2m_1^2}{t_2^2s^2}$

$$h_2 = \sqrt{\frac{(2B + AR)}{-A}B + d}$$

$$\cos\beta_R = \frac{1}{Rv_R}\sqrt{\frac{(2B + AR)}{-A}B + d}$$

Approximating the expression in the first radical sign with Umer Khayams's binomial series expansion *before* combining the two terms, we finally obtain as below:

For $0 < v_2 <<< \sqrt{\frac{2Gm_1}{t_2^2R}} \Rightarrow \beta_2 \approx 45$ degrees.

Where $t_2 = 1 + \frac{m_2}{m_1}$

For $v_2 = \sqrt{\frac{Gm_1}{t_2^2R}} \Rightarrow \beta_2 \approx 0$ degrees.

For velocities in the range as below

$$\sqrt{\frac{Gm_2}{t_1^2r_1(1-3\frac{Gm_2}{r_1s^2})}} \leq v < \frac{s}{t_1}\sqrt{\frac{1}{3}(e^{\frac{6Gm_2}{s^2r_1}} - 1)} \quad \text{exact}$$

$$\sqrt{\frac{Gm_1}{t_2^2R}} \leq v < \sqrt{\frac{2Gm_1}{t_2^2R}} \quad \text{approx.}$$

β_R stays approximately equal to zero i.e., $\beta_R \approx 0$ for maximum range! The only difference is that the orbits are non-circular now. This is also popular among aliens as a source of recreational balls. In the *approximately* elliptical, parabolic and hyperbolic scenarios the body doesn't fall back on earth because of which the scenarios are not covered by the Tahirian equation.

Now let me produce, for the first and the last Time in History, the actual Equation of Gravitation, i.e., The Tahirian Equation of Gravitation – The Superior Equation of Tahir, that eventually connects the motion of the Tahir's Rising Apple- Tahir's Projectiles- with the motion of planets and then of Binary stars. But to proceed we need to produce what no brains have yet thought of, i.e., the Tahir's Field Equations (TFEs) as below:

$$\frac{dr}{dt} = v\sin\beta \quad \text{(Tahir's Partial Field Equation-One) (TPFE-1)}$$

$$r\frac{d\theta}{dt} = v\cos\beta \quad \text{(Tahir's Partial Field Equation-Two) (TPFE-2)}$$

The above is not partial in the sense of partial differential equations! It just means pieces/parts of a bigger thing.

Dividing TPFE-1 by the TPFE-2 we get the TFE as below:

$$\frac{dr}{rd\theta} = \tan\beta$$

$$d\theta = \frac{dr}{r\tan\beta}$$

$$\theta = \int \frac{dr}{r\tan\beta}$$

(Tahir Field equation (TFE))

Constant of integration is intentionally chosen to be equal to the weight of current relativistic theories! When we solve the integral, we will see that the choice was correct.

Now let Tahir do the barbeque of the above TFE as below:

$$\theta = \int \frac{dr}{r\tan\beta} = \int \frac{dr}{r\sqrt{\sec^2\beta - 1}}$$

$$\theta = \int \frac{dr}{r\sqrt{\frac{1}{\cos^2\beta} - 1}} = \int \frac{dr}{r\sqrt{\frac{r^2v^2}{r^2v^2\cos^2\beta} - 1}} = \int \frac{dr}{r\sqrt{\frac{r^2v^2}{h^2} - 1}}$$

$$\theta = \int \frac{dr}{r\sqrt{\frac{r^2v^2}{h^2} - 1}} = \int \frac{hdr}{r\sqrt{r^2v^2 - h^2}}$$

Let us represent $\frac{1}{r}$ by u. Therefore, we have:

$$u = \frac{1}{r} \quad \text{and} \quad du = -\frac{dr}{r^2} \quad \text{or} \quad \frac{du}{u} = -\frac{dr}{r}$$

$$\theta = \int \frac{hdr}{r\sqrt{r^2v^2 - h^2}} = -\int \frac{hdu}{u\sqrt{\frac{v^2}{u^2} - h^2}} = -\int \frac{hdu}{\sqrt{v^2 - h^2u^2}}$$

$$\boxed{\theta = -\int \frac{hdu}{\sqrt{v^2 - h^2u^2}} = -h\int \frac{du}{\sqrt{au^2 + bu + c}}}$$

Now, as has already gone above, we have:

$$V^2 - V_R^2 = 2GM(\frac{1}{r} - \frac{1}{R})$$

$$V^2 = V_R^2 + 2GM(\frac{1}{r} - \frac{1}{R})$$

Putting the above value into the integral we obtain for non-gravitivistic results of static orbits, the following integral:

$$\theta = -\int \frac{hdu}{\sqrt{v^2 - h^2u^2}} = -\int \frac{hdu}{\sqrt{V_R^2 + 2GM(\frac{1}{r} - \frac{1}{R}) - h^2u^2}}$$

$$\theta = -\int \frac{hdu}{\sqrt{V_R^2 + 2GM(u - \frac{1}{R}) - h^2u^2}} = -h\int \frac{du}{\sqrt{V_R^2 - \frac{2GM}{R} + 2GMu - h^2u^2}}$$

Now from [12]

$$\int \frac{du}{\sqrt{au^2 + bu + c}} = -\frac{1}{\sqrt{-a}} \arcsin \frac{2au + b}{\sqrt{-\Delta}} \quad \{a < 0, \Delta < 0,\}$$

$$\Delta = 4ac - b^2$$

$$\theta = \frac{h}{\sqrt{-a}} \arcsin \frac{2au + b}{\sqrt{-\Delta}}$$

For our case, we have

$$a = -h^2$$

$$b = 2GM$$

$$c = V_R^2 - \frac{2GM}{R}$$

$$\boxed{\theta = -\int \frac{h\,du}{\sqrt{v^2 - h^2 u^2}} = -h \int \frac{du}{\sqrt{au^2 + bu + c}}}$$

Inserting the values, we obtain as below:

$$\theta = -h \times -\frac{1}{\sqrt{h^2}} \arcsin \frac{-2h^2 u + 2GM}{\sqrt{(2GM)^2 + 4h^2(V_R^2 - \frac{2GM}{R})}}$$

$$\sin\theta = \frac{-2h^2 u + 2GM}{\sqrt{4G^2M^2 + 4h^2(V_R^2 - \frac{2GM}{R})}}$$

$$\sin\theta = \frac{-2h^2 u + 2GM}{2h\sqrt{\frac{G^2M^2}{h^2} + (V_R^2 - \frac{2GM}{R})}}$$

$$\sin\theta = \frac{-hu + \frac{GM}{h}}{\sqrt{\frac{G^2M^2}{h^2} + (V_R^2 - \frac{2GM}{R})}}$$

$$\sqrt{\frac{G^2M^2}{h^2} + (V_R^2 - \frac{2GM}{R})}\sin\theta = -hu + \frac{GM}{h}$$

$$hu = \frac{GM}{h} - \sqrt{\frac{G^2M^2}{h^2} + (V_R^2 - \frac{2GM}{R})}\sin\theta$$

$$\frac{h}{r} = \frac{GM}{h} - \sqrt{\frac{G^2M^2}{h^2} + (V_R^2 - \frac{2GM}{R})}\sin\theta$$

$$r = \frac{h}{\frac{GM}{h} - \sqrt{\frac{G^2M^2}{h^2} + (V_R^2 - \frac{2GM}{R})}\sin\theta}$$

$$r = \frac{h}{\frac{GM}{h}\left(1 - \frac{h}{GM}\sqrt{\frac{G^2M^2}{h^2} + \left(V_R^2 - \frac{2GM}{R}\right)}\sin\theta\right)}$$

$$r = \frac{h^2/GM}{1 - \sqrt{1 + \frac{h^2}{G^2M^2}\left(V_R^2 - \frac{2GM}{R}\right)}\sin\theta}$$

$$r = \frac{h^2/GM}{1 - \varepsilon\sin\theta}$$

Where:

$$\varepsilon = \sqrt{1 + \frac{h^2}{G^2M^2}\left(V_R^2 - \frac{2GM}{R}\right)} = \text{Eccentricity of the ellipse}$$

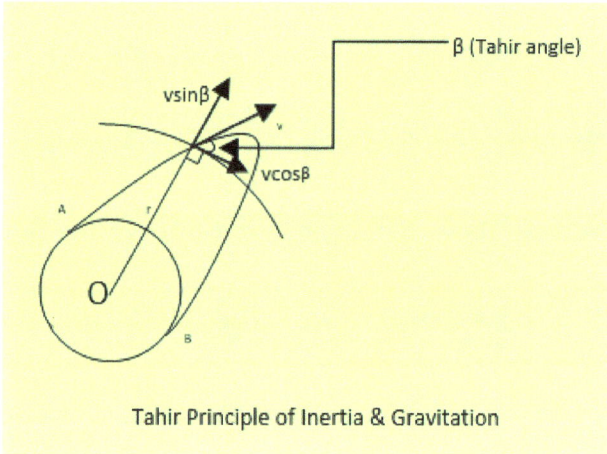

Tahir Principle of Inertia & Gravitation

Figure 20. Tahirian trajectories of projectiles

Now let me find the optimal value of the Tahir Angle at which the projectile reaches its maximum distance from the point of projection on the surface of the earth once again but this time using a single independent variable. Previously Dr. Tahir used two independent variables. Why I chose zero as the constant of integration is now evident from the above figure because the zero value of the constant of integration gives a good orientation for us to draw the elliptical Trajectory of Tahirian non-gravitivistic Projectile as above in the figure for the given set of x, y and polar axes.

From the above figure, we have

$$\theta_0 = \pi - 2\psi$$

So, the distance travelled by the projectile is the arc of the circle of earth as below:

$$S = R\theta_0 = R(\pi - 2\psi)$$

We can rewrite the above equation of trajectories a little bit differently as below:

$$r = \frac{h^2/GM}{1 - \sqrt{1 + \frac{h^2}{G^2M^2}\left(V_R^2 - \frac{2GM}{R}\right)}\sin\theta}$$

$$r = \frac{h}{\frac{GM}{h} - \sqrt{\frac{G^2M^2}{h^2} + \left(V_R^2 - \frac{2GM}{R}\right)}\sin\theta}$$

or

Natural Theory of Relativity, Inertia, Gravitation & Gravitivity

$$r = \frac{h}{\frac{GM}{h} - \sqrt{D}\sin\theta}$$

Where $D = \frac{G^2M^2}{h^2} + \left(v_R^2 - \frac{2GM}{R}\right) = \frac{b^2}{-4a} + c = \frac{b^2 - 4ac}{-4a} = \frac{\Delta}{4a}$

Putting the value of $\theta = \psi$ in the equation of static trajectories of projectiles and planets, we obtain as below:

$$R = \frac{h}{\frac{GM}{h} - \sqrt{D}\sin\psi}$$

We have

$$\psi = \arcsin\left(\frac{\frac{GM}{h} - \frac{h}{R}}{\sqrt{D}}\right)$$

$$\psi = \arcsin\left(\frac{\frac{GM}{h} - V_R\cos\beta_R}{\sqrt{D}}\right)$$

Putting the above value in the previous angle equation, we obtain as below:

$$S = R\theta_0 = R(\pi - 2\psi)$$

$$S = R\theta_0 = \pi R - 2R\psi$$

$$\psi \leq \theta \leq \psi + \theta_0$$

$$S = \pi R - 2R\arcsin\left(\frac{\frac{GM}{h} - V_R\cos\beta_R}{\sqrt{D}}\right)$$

Differentiating the above equation with respect to β_R, we obtain

$$\frac{dS}{d\beta_R} = -2R\frac{d\psi}{d\beta_R} = -2R\frac{d}{d\beta_R}\arcsin\left(\frac{\frac{GM}{h} - V_R\cos\beta_R}{\sqrt{D}}\right)$$

$$\frac{dS}{d\beta_R} = \frac{-2R}{\sqrt{1 - \left(\frac{\frac{GM}{h} - V_R\cos\beta_R}{\sqrt{D}}\right)^2}}\frac{d}{d\beta_R}\left(\frac{\frac{GM}{h} - V_R\cos\beta_R}{\sqrt{D}}\right)$$

Setting $\frac{dS}{d\beta_R} = 0$ for the optimal value of β_R in the above equation, Tahir obtains as below:

$$\frac{d}{d\beta_R}\left(\frac{\frac{GM}{h} - V_R\cos\beta_R}{\sqrt{D}}\right) = 0$$

We can write the above equation as

$$\frac{d}{d\beta_R}\left(\frac{\frac{GM}{h} - V_R\cos\beta_R}{\sqrt{D}}\right) = 0$$

$$D = \frac{G^2M^2}{h^2} + \left(V_R^2 - \frac{2GM}{R}\right) = \frac{B^2}{h^2} + A$$

Where $A = v_R^2 - 2\frac{GM}{R}$, $B = GM$ and $h_R = RV_R\cos\beta_R$

And $\frac{dh}{d\beta_R} \neq 0$

$$\left(\sqrt{\frac{B^2}{h^2} + A}\right)\frac{d}{d\beta_R}\left(\frac{GM}{h} - V_R\cos\beta_R\right) - \left(\frac{GM}{h} - V_R\cos\beta_R\right)\frac{d}{d\beta_R}\left(\sqrt{\frac{B^2}{h^2} + A}\right) = 0$$

$$\left(\sqrt{\frac{B^2}{h^2} + A}\right)\frac{d}{d\beta_R}\left(\frac{B}{h} - V_R\cos\beta_R\right) - \left(\frac{B}{h} - V_R\cos\beta_R\right)\frac{d}{d\beta_R}\left(\sqrt{\frac{B^2}{h^2} + A}\right) = 0$$

$$\left(\sqrt{\frac{B^2}{h^2} + A}\right)\frac{d}{d\beta_R}\left(\frac{B}{RV_R\cos\beta_R} - V_R\cos\beta_R\right) - \left(\frac{B}{RV_R\cos\beta_R} - V_R\cos\beta_R\right)\frac{d}{d\beta_R}\left(\sqrt{\frac{B^2}{h^2} + A}\right) = 0$$

V_R must be kept constant in the above differentiation process.

After doing some exercise in calculus, Tahir obtains as before.

$$p = \frac{(2B + AR)}{-A}B$$

Where:

$$h^2 = p$$

$$h = \sqrt{\frac{(2B + AR)}{-A}B}$$

Hence, we obtain as below:

$$\cos\beta_R = \frac{1}{Rv_R}\sqrt{\left[\frac{1}{1 - \frac{v_R^2 R}{2GM}} - 1\right]GMR}$$

The above result is remarkable and the same as got before by Tahir for the first Time in History once and forever by using partial differentiation and the rule of parallel vectors. The alien method is a third-world method and hence fails to produce the Superior First World Tahirian Gravitivistic Value(TGV) of the optimal angle of launch. For your information, the alien method has not yet been able to develop even faith in the existence of a Tahirian non-Gravitivistic value!

The above result can also be got relatively easily by converting the quantity $V_R\cos\beta_R$ in to h_R and differentiating w.r.t h_R as below:

$$\frac{d}{dh_R}\left(\frac{\frac{GM}{h} - \frac{h_R}{R}}{\sqrt{D}}\right) = 0$$

Where $D = \frac{G^2M^2}{h^2} + \left(V_R^2 - \frac{2GM}{R}\right)$

And $h = h_R$

$$\frac{d}{dh_R}\left(\frac{\frac{GM}{h_R} - \frac{h_R}{R}}{\sqrt{\frac{G^2M^2}{h_R^2} + \left(V_R^2 - \frac{2GM}{R}\right)}}\right) = 0$$

Or

$$\frac{d}{dh_R}\left(\frac{\dfrac{B}{h_R}-\dfrac{h_R}{R}}{\sqrt{\dfrac{B^2}{h_R^{2}}+A}}\right)=0$$

A, B and R should be kept constant throughout the process of differentiation. Differentiation is left as an exercise. There is no book in the world that contains such an exercise other than Kitab-e-Tahir!

Hence, we obtain as before:

$$h=\sqrt{\frac{(2B+AR)}{-A}}\,B$$

For the gravitivistic value of the optimal angle of launch we can proceed as below

$$\theta=-\int\frac{h\,du}{\sqrt{v^2-h^2u^2}}=-h\int\frac{du}{\sqrt{au^2+bu+c}}$$

$$\int\frac{du}{\sqrt{au^2+bu+c}}=-\frac{1}{\sqrt{-a}}\arcsin\frac{2au+b}{\sqrt{-\Delta}}\quad\{a<0,\Delta<0,\}$$

$$\theta=-\int\frac{h\,du}{\sqrt{v^2-h^2u^2}}=-h\times-\frac{1}{\sqrt{-a}}\arcsin\frac{2au+b}{\sqrt{-\Delta}}$$

$$\theta=\frac{h}{\sqrt{-a}}\arcsin\frac{2au+b}{\sqrt{b^2-4ac}}$$

When $\theta=\psi$; $u=1/R$

$$S=R\theta_0=R(\pi-2\psi)$$

$$a=\frac{6G^2m_1^2}{t_2^2s^2}-h_2^2=d-h_2^2$$

$$b=\frac{2Gm_1}{t_2^2}-\frac{12G^2m_1^2}{t_2^2s^2r_{21}}+\frac{6Gm_1v_{21}^2}{s^2}$$

$$c=v_{21}^2-\frac{2Gm_1}{t_2^2r_{21}}+\frac{6G^2m_1^2}{t_2^2s^2r_{21}^2}-\frac{6Gm_1v_{21}^2}{s^2r_{21}}$$

$$A=v_{2R}^2-\frac{2Gm_1}{t_2^2r_{2R}}-\frac{6Gm_1v_{21}^2}{s^2r_{2R}}+\frac{6G^2m_1^2}{t_2^2s^2r_{2R}^2}$$

$$B=\frac{Gm_1}{t_2^2}-\frac{6G^2m_1^2}{t_2^2s^2r_{2R}}+\frac{3Gm_1v_{21}^2}{s^2}$$

$$C=\frac{6G^2m_1^2}{t_2^2s^2}-h_2^2$$

For the values of a, b & c see next page.

$$S=R\theta_0=\pi R-2R\frac{h}{\sqrt{-a}}\arcsin\frac{(\frac{2a}{R}+b)}{\sqrt{b^2-4ac}}$$

$$\frac{d\left[\frac{h}{\sqrt{-a}}\arcsin\dfrac{(\frac{2a}{R}+b)}{\sqrt{b^2-4ac}}\right]}{da}=0$$

Here I can play a trick to make life simple as below. We will write h as below:

$$\frac{d\left[\sqrt{\dfrac{h^2}{-a}}\arcsin\dfrac{(\frac{2a}{R}+b)}{\sqrt{b^2-4ac}}\right]}{da}=0$$

$$\frac{d\left[\sqrt{\dfrac{d-a}{-a}}\arcsin\dfrac{(\frac{2a}{R}+b)}{\sqrt{b^2-4ac}}\right]}{da}=0$$

$$\frac{d\left[\sqrt{1-\dfrac{d}{a}}\arcsin\dfrac{(\frac{2a}{R}+b)}{\sqrt{b^2-4ac}}\right]}{da}=0$$

b,c & d should be considered constant during the differentiation process because they are either constants or initial constant conditions in the trajectory of the projectile.

So, after wrestling differentially we come to the juncture during the process of simplification as below:

$$\frac{d}{2a}\arcsin\frac{(\frac{2a}{R}+b)}{\sqrt{b^2-4ac}}=\frac{(a-d)(\frac{2b^2}{R}-\frac{4ac}{R}+2bc)}{\sqrt{b^2-4ac-\left(\frac{2a}{R}+b\right)}\,(b^2-4ac)}$$

The above equation *can't* be solved exactly- up to powers of s^2 or less - unless or until we square both sides of the equation to get rid of the square root sign in the denominator because it contains terms containing a! When we square both sides, d gets squared and gives us a term containing s^4 in the denominator which makes it extremely small and hence not required by Tahir. So, we are simply left with the equation as below:

$$\frac{2b^2}{R}-\frac{4ac}{R}+2bc=0$$

$$\frac{4ac}{R}=\frac{2b^2}{R}+2bc$$

$$2ac=b^2+bcR$$

$$a=\frac{b^2+bcR}{2c}=d-h^2$$

$$\frac{b^2+bcR}{2c}=d-h^2$$

$$h^2=\frac{b^2+bcR}{-2c}+d$$

$$h=\sqrt{\frac{(b+cR)}{-2c}b+d}$$

Now I know from previous work as below:

$$a=C$$

$$\mathbf{b = 2B}$$

$$\mathbf{c = A}$$

$$\mathbf{h} = \sqrt{\frac{(2B + AR)}{-A}\,B + d}$$

Now let us use the following equation to find the Tahirian Precession of the planets under binary star system with the sun.

$$v_{22}^2 = \frac{s^2}{3t_2^2}\left\{\left(1 + 3t_2^2\frac{v_{21}^2}{s^2}\right)e^{\left[\frac{6Gm_1}{s^2}\left(\frac{1}{r_{22}} - \frac{1}{r_{21}}\right)\right]} - 1\right\}$$

But we first need to expand the equation to reasonable accuracy in the direction of Natural Tahirian Philosophy (NTP) as below:

$$v_{22}^2 = \frac{s^2}{3t_2^2}\left\{\left(1 + 3t_2^2\frac{v_{21}^2}{s^2}\right)\left(1 + \frac{6Gm_1}{s^2}\left(\frac{1}{r_{22}} - \frac{1}{r_{21}}\right)\right.\right.$$
$$\left.\left. + \frac{1}{2!}\left(\frac{6Gm_1}{s^2}\left(\frac{1}{r_{22}} - \frac{1}{r_{21}}\right)\right)^2 + \cdots\right) - 1\right\}$$

Expanding the above we obtain as below:

$$v_{22}^2 = v_{21}^2 - \frac{2Gm_1}{t_2^2 r_{21}} + \frac{2Gm_1}{t_2^2 r_{22}} + \frac{6G^2m_1^2}{t_2^2 s^2 r_{22}^2} - \frac{12G^2m_1^2}{t_2^2 s^2 r_{21} r_{22}}$$
$$+ \frac{6G^2m_1^2}{t_2^2 s^2 r_{21}^2} + \frac{6Gm_1 v_{21}^2}{s^2 r_{22}} - \frac{6Gm_1 v_{21}^2}{s^2 r_{21}} + \cdots$$

We have already represented $\frac{1}{r}$ by u. Therefore we have:

$$v_{22}^2 = v_{21}^2 - \frac{2Gm_1}{t_2^2 r_{21}} + \frac{2Gm_1}{t_2^2}u_{22} + \frac{6G^2m_1^2}{t_2^2 s^2}u_{22}^2$$
$$- \frac{12G^2m_1^2}{t_2^2 s^2 r_{21}}u_{22} + \frac{6G^2m_1^2}{t_2^2 s^2 r_{21}^2} + \frac{6Gm_1 v_{21}^2}{s^2}u_{22}$$
$$- \frac{6Gm_1 v_{21}^2}{s^2 r_{21}} + \cdots$$

$$v_{22}^2 = v_{21}^2 - \frac{2Gm_1}{t_2^2 r_{21}} + \frac{6G^2m_1^2}{t_2^2 s^2 r_{21}^2} - \frac{6Gm_1 v_{21}^2}{s^2 r_{21}} + \frac{2Gm_1}{t_2^2}u_{22}$$
$$- \frac{12G^2m_1^2}{t_2^2 s^2 r_{21}}u_{22} + \frac{6Gm_1 v_{21}^2}{s^2}u_{22}$$
$$+ \frac{6G^2m_1^2}{t_2^2 s^2}u_{22}^2 + \cdots$$

$$v_{22}^2 = \left(v_{21}^2 - \frac{2Gm_1}{t_2^2 r_{21}} + \frac{6G^2m_1^2}{t_2^2 s^2 r_{21}^2} - \frac{6Gm_1 v_{21}^2}{s^2 r_{21}}\right) + \left(\frac{2Gm_1}{t_2^2}\right.$$
$$\left. - \frac{12G^2m_1^2}{t_2^2 s^2 r_{21}} + \frac{6Gm_1 v_{21}^2}{s^2}\right)u_{22}$$
$$+ \frac{6G^2m_1^2}{t_2^2 s^2}u_{22}^2 + \cdots$$

Now, we can easily assume that the origin of coordinates related to TFEs coincides with the Tahir Centre of rotation TCR of the binary. Therefore, putting the above value into the equation below and as went before, we obtain as below:

$$\theta_{22} = -\int \frac{h_2 du_{22}}{\sqrt{v_{22}^2 - h_2^2 u_{22}^2}}$$

$$v_{22}^2 - h_2^2 u_{22}^2 = \left(v_{21}^2 - \frac{2Gm_1}{t_2^2 r_{21}} + \frac{6G^2m_1^2}{t_2^2 s^2 r_{21}^2} - \frac{6Gm_1 v_{21}^2}{s^2 r_{21}}\right)$$
$$+ \left(\frac{2Gm_1}{t_2^2} - \frac{12G^2m_1^2}{t_2^2 s^2 r_{21}} + \frac{6Gm_1 v_{21}^2}{s^2}\right)u_{22}$$
$$+ \left(\frac{6G^2m_1^2}{t_2^2 s^2} - h_2^2\right)u_{22}^2 + \cdots$$

$$v_{22}^2 - h_2^2 u_{22}^2 \approx \left(\frac{6G^2m_1^2}{t_2^2 s^2} - h_2^2\right)u_{22}^2$$
$$+ \left(\frac{2Gm_1}{t_2^2} - \frac{12G^2m_1^2}{t_2^2 s^2 r_{21}} + \frac{6Gm_1 v_{21}^2}{s^2}\right)u_{22}$$
$$+ \left(v_{21}^2 - \frac{2Gm_1}{t_2^2 r_{21}} + \frac{6G^2m_1^2}{t_2^2 s^2 r_{21}^2} - \frac{6Gm_1 v_{21}^2}{s^2 r_{21}}\right)$$

$$v_{22}^2 - h_2^2 u_{22}^2 \approx au_{22}^2 + bu_{22} + c$$

$$a = \frac{6G^2m_1^2}{t_2^2 s^2} - h_2^2$$

$$b = \frac{2Gm_1}{t_2^2} - \frac{12G^2m_1^2}{t_2^2 s^2 r_{21}} + \frac{6Gm_1 v_{21}^2}{s^2}$$

$$c = v_{21}^2 - \frac{2Gm_1}{t_2^2 r_{21}} + \frac{6G^2m_1^2}{t_2^2 s^2 r_{21}^2} - \frac{6Gm_1 v_{21}^2}{s^2 r_{21}}$$

For precession of the planets, we need to consider the coefficient **a** only as below

$$\int \frac{du}{\sqrt{au^2 + bu + c}} = -\frac{1}{\sqrt{-a}}\arcsin\frac{2au + b}{\sqrt{b^2 - 4ac}} \quad (\text{for } a < 0, b^2 - 4ac > 0, |2au + b| < \sqrt{b^2 - 4ac})$$

$$\theta_{22} = -\int \frac{h_2 du_{22}}{\sqrt{v_{22}^2 - h_2^2 u_{22}^2}} = -h_2\int \frac{h_2 du_{22}}{\sqrt{v_{22}^2 - h_2^2 u_{22}^2}} = -h_2\int \frac{du_{22}}{\sqrt{au_{22}^2 + bu_{22} + c}}$$

$$\theta_{22} = -h_2\int \frac{du_{22}}{\sqrt{au_{22}^2 + bu_{22} + c}}$$

$$\theta_{22} = -h_2 \times -\frac{1}{\sqrt{-a}}\arcsin\frac{2au + b}{\sqrt{b^2 - 4ac}}$$

$$\theta_{22} = -h_2 \times -\frac{1}{\sqrt{h_2^2 - \frac{6G^2m_1^2}{t_2^2 s^2}}}\arcsin\frac{2au + b}{\sqrt{b^2 - 4ac}}$$

$$\theta_{22} = \frac{1}{\sqrt{1 - \frac{6G^2m_1^2}{t_2^2 h_2^2 s^2}}}\arcsin\frac{2au + b}{\sqrt{b^2 - 4ac}}$$

$$\left[\left\{\sqrt{1 - \frac{6G^2m_1^2}{t_2^2 h_2^2 s^2}}\right\}\theta_{22}\right] = \arcsin\frac{2au + b}{\sqrt{b^2 - 4ac}}$$

$$\sin\left[\left\{\sqrt{1 - \frac{6G^2m_1^2}{t_2^2 h_2^2 s^2}}\right\}\theta_{22}\right] = \frac{2au + b}{\sqrt{b^2 - 4ac}}$$

$$\sin\left[\left\{1 - \frac{3G^2m_1^2}{t_2^2 h_2^2 s^2}\right\}\theta_{22}\right] \approx \frac{2au + b}{\sqrt{b^2 - 4ac}}$$

$$\sin\left[\left\{1 - \frac{3G^2m_1^2}{t_2^2 h_2^2 s^2}\right\}\theta_{22}\right] \approx \frac{2au_{22} + b}{\sqrt{b^2 - 4ac}}$$

When $\theta_{22} = 2\pi$, we obtain a total gravitivistic precession of the planet, under binary system with the sun or any other planet, in the forward direction of motion of the magnitude of

$$2\pi \times \frac{3G^2m_1^2}{t_2^2 h_2^2 s^2} = 6\pi\frac{G^2m_1^2}{t_2^2 h_2^2 s^2} \text{ radians per revolution}$$

We see that Dr. Tahir has been able to correct the term for binary star system by adding the Tahir Binary Star System Constant term $\mathbf{t_2^2}$. If $h_1 = h_2$ then the precessions of both the sun and the planet will be equal and have the magnitude four times less than the magnitude otherwise. For the current configuration of the Sun and the planets its value is very close to 1. I think that in Tahirian Cosmos it is not possible for two particles of matter to orbit around each other in perfect circles under binary system and, of-course, all systems are at least binaries. Why? Because, in that scenario of both the particles circling in perfect circular orbits, it negates the concept of orbital precession which is evident only under elliptical orbits!

Solving the equation below for u_{22}, we obtain

$$\sin[\{\sqrt{1 - \frac{6G^2 m_1^2}{t_2^2 h_2^2 s^2}}\}\theta_{22}] \approx \frac{2a u_{22} + b}{\sqrt{b^2 - 4ac}}$$

$$2a u_{22} + b \approx \sqrt{b^2 - 4ac}\sin[\{\sqrt{1 - \frac{6G^2 m_1^2}{t_2^2 h_2^2 s^2}}\}\theta_{22}]$$

$$u_{22} \approx \frac{\sqrt{b^2 - 4ac}\sin[\{\sqrt{1 - \frac{6G^2 m_1^2}{t_2^2 h_2^2 s^2}}\}\theta_{22}] - b}{2a}$$

$$\frac{1}{r_{22}} \approx \frac{\sqrt{b^2 - 4ac}\sin[\{\sqrt{1 - \frac{6G^2 m_1^2}{t_2^2 h_2^2 s^2}}\}\theta_{22}] - b}{2a}$$

$$r_{22} \approx \frac{2a}{\sqrt{b^2 - 4ac}\sin[\{\sqrt{1 - \frac{6G^2 m_1^2}{t_2^2 h_2^2 s^2}}\}\theta_{22}] - b}$$

$$r_{22} \approx \frac{2a}{\sqrt{b^2 - 4ac}\sin[\{\sqrt{1 - \frac{6G^2 m_1^2}{t_2^2 h_2^2 s^2}}\}\theta_{22}] - b}$$

$$r_{22} \approx \frac{2a}{-b + \sqrt{b^2 - 4ac}\sin[\{\sqrt{1 - \frac{6G^2 m_1^2}{t_2^2 h_2^2 s^2}}\}\theta_{22}]}$$

Or simply

$$r_{22} \approx \frac{-2a}{b - \sqrt{b^2 - 4ac}\sin[\{\sqrt{1 - \frac{6G^2 m_1^2}{t_2^2 h_2^2 s^2}}\}\theta_{22}]}$$

$$r_{22} \approx \frac{-2a/b}{1 - \frac{\sqrt{b^2 - 4ac}}{b}\sin[\{\sqrt{1 - \frac{6G^2 m_1^2}{t_2^2 h_2^2 s^2}}\}\theta_{22}]}$$

I think it is not possible for the human brain to produce something better than the above in the next 4180 years at least!

Another interesting form can also be written as below:

$$r_{22} \approx \frac{1}{\dfrac{-b + \sqrt{b^2 - 4ac}\sin[\{\sqrt{1 - \frac{6G^2 m_1^2}{t_2^2 h_2^2 s^2}}\}\theta_{22}]}{2a}}$$

It looks like the reciprocal of one of the Alkhwarizmic roots of the quadratic equation!

$$\varepsilon = \sqrt{1 + \frac{h^2}{G^2 M^2}\left(V_R^2 - \frac{2GM}{R}\right)} \quad \text{(Non-gravitivistic)}$$

$$\varepsilon = \frac{\sqrt{b^2 - 4ac}}{b} \quad \text{(Gravitivistic)}$$

Theoretically, we can't say that an eccentricity exists for gravitivistic orbits because they are not closed elliptical orbits but still very much like them!

In short, my technique is much better than that of noton, ooler, langralagian, heemama-linimilton etc. or whoever alien enquired into the field of Islamic Celestial Mechanics.

6. Double deflection of photons

$$F_t = \left(1 + 3\frac{v^2}{c^2}\right)\frac{Gm_1 m_2}{r^2}\sin\beta = \frac{dV}{dT}$$

The above force F_t is the rectangular component of the gravitational force acting on the test particle (when going away from the gravitating mass) *against* the direction of motion or tangential to the direction of motion and the *inactive* component normal to is as below:

$$F_n = \frac{Gm_1 m_2}{r^2}\left(1 + 3\frac{v^2}{c^2}\right)\cos\beta$$

The resultant gravitational force is given by the magnitude of the rectangular components as below by Alkashi's Theorem in the rectangular case.

$$F_{Total}^2 = F_{Tangential}^2 + F_{Normal}^2$$

$$F_{Total}^2 = \sqrt{[\left(1 + 3\frac{v^2}{c^2}\right)\frac{Gm_1 m_2}{r^2}\sin\beta]^2 + [\frac{Gm_1 m_2}{r^2}\left(1 + 3\frac{v^2}{c^2}\right)\cos\beta]^2}$$

$$F_{total} = \left(1 + 3\frac{v^2}{c^2}\right)\frac{Gm_1 m_2}{r^2}$$

Diagrams can be drawn. They are very simple. The tangential component is negative when the test particle is going away from the gravitating mass and vice versa. The rectangular components are such that the resultant/total gravitational force is always ACTING towards the gravitating mass. Amazing, isn't it? The net force $m_1 \frac{dV_1}{dT}$ is the projected resultant (or simply the resultant when the payload is released) which in turn is the active component of gravity $-\left(1 + 3\frac{v^2}{c^2}\right)\frac{Gm_1 m_2}{r^2}\sin\beta$.

$-\frac{Gm_1 m_2}{r^2}\left(1 + 3\frac{v^2}{c^2}\right)\cos\beta$ is the inactive component of gravity. Inactive doesn't mean that it has lost the status of being the *component* of gravity!

Now let us derive the same result by using the symbols of Al-jebra used for recreation by the aliens. The natural differential line element is given as

$$dS = v'dT'$$

Squaring we obtain

$$dS^2 = \left(\frac{v}{(1-v^2/c^2)}\right)^2 dT'^2$$

Let $\psi = 1 - v^2/c^2$

$$dS'^2 = \frac{v^2}{\psi^2} dT'^2 = g_{pq} dx^p dx^q$$

The Tahir Single Element Metric above yields

$$g_{pq} = \left[\frac{v^2}{\psi^2}\right] = g_{11}$$

Determinant is used to find the inverse of a matrix, if it exists, only when the order of the matrix is either 2×2 or greater. When the matrix is 1×1 then we don't need determinant to find the inverse of a matrix. The inverse then is just the reciprocal of the single element or number. So, in line with what has been said we have:

$$g^{pq} = \frac{1}{g_{pq}} = \frac{1}{\left[\frac{v^2}{\psi^2}\right]} = \left[\frac{v^2}{\psi^2}\right]^{-1} = \left[\frac{\psi^2}{v^2}\right] = g^{11}$$

The contravariant component of acceleration in our case is nothing but the geodesic equation as below [4]

$$a^k = \frac{d^2 x^s}{d\lambda^2} + \Gamma_{pq}^s \frac{dx^p}{d\lambda} \frac{dx^q}{d\lambda} = 0$$

Where:

$$\Gamma_{pq}^s = \frac{1}{2} g^{sr}\left(\frac{\partial g_{pr}}{\partial x^q} + \frac{\partial g_{qr}}{\partial x^p} - \frac{\partial g_{pq}}{\partial x^r}\right)$$

The physical component of the acceleration is given below as

$$a_{physical} = \sqrt{g_{11}} a^k = \sqrt{g_{11}}\left[\frac{d^2 x^s}{d\lambda^2} + \Gamma_{pq}^s \frac{dx^p}{d\lambda} \frac{dx^q}{d\lambda}\right] = 0$$

Now solving the *only* alien symbol for our *Tahir Single Element Metric*, we obtain

$$\Gamma_{11}^1 = \frac{1}{2} g^{11}\left(\frac{\partial g_{11}}{\partial x^1} + \frac{\partial g_{11}}{\partial x^1} - \frac{\partial g_{11}}{\partial x^1}\right)$$

$$\Gamma_{11}^1 = \frac{1}{2} g^{11} \frac{\partial g_{11}}{\partial x^1} = \frac{1}{2g_{11}} \frac{\partial g_{11}}{\partial x^1}$$

$$\Gamma_{11}^1 = \frac{1}{2g_{11}} \frac{\partial g_{11}}{\partial x^1} = \frac{\psi^2}{2v^2} \frac{\partial}{\partial x^1}\left(\frac{v^2}{\psi^2}\right)$$

$$\Gamma_{11}^1 = \frac{\psi^2}{2v^2} \times 2\frac{v}{\psi} \times \frac{\partial}{\partial x^1}\left(\frac{v}{\psi}\right)$$

$$\Gamma_{11}^1 = \frac{\psi}{v} \times \frac{\partial}{\partial x^1}\left(\frac{v}{1-v^2/c^2}\right)$$

Other recreational symbols in Al-jebra don't exist because there is only one element of Tahir Single Metric/Matrix. Zeros as another option for elements can't be construed as such because it would render the Matrix invalid for inversion by making determinant equal to zero. Determinant is not the vital factor in our case because we can find the inverse simply by taking the reciprocal of the single element of the single matrix/metric. Nevertheless, determinant does help in understanding the non-existence of other elements. Matrix Theory of 1×1 matrix considers zero as a number *only* whereas ordinarily zero is a number and the absence of a number at the same time. For example, in the number 2019, zero can't be ignored because it represents the absence of a number from 1,2,3,4,5,6,7,8 & 9. Otherwise, it would become 219 if we don't acknowledge the existence of zero in 2019 as to represent the absence of a number and a number at the same time.

Where $x^1 = T'$, we now have:

$$\Gamma_{11}^1 = \frac{\psi}{v} \times \frac{\partial}{\partial T'}\left(\frac{v}{1-v^2/c^2}\right)$$

We see above that the function to be differentiated is a function of *one* variable *only* namely v. So, we can write as below:

$$\frac{\partial}{\partial T'}\left(\frac{v}{1-v^2/c^2}\right) = \frac{d}{dT'}\left(\frac{v}{1-v^2/c^2}\right)$$

$$\Gamma_{11}^1 = \frac{\psi}{v} \times \frac{\partial}{\partial T'}\left(\frac{v}{1-v^2/c^2}\right) = \frac{\psi}{v} \times \frac{d}{dT'}\left(\frac{v}{1-v^2/c^2}\right)$$

Now applying the chain rule, we finally obtain

$$\Gamma_{11}^1 = \frac{\psi}{v} \times \frac{\partial}{\partial T'}\left(\frac{v}{1-v^2/c^2}\right) = \frac{\psi}{v} \times \frac{d}{dv}\left(\frac{v}{1-v^2/c^2}\right)\frac{dv}{dT'}$$

$$\Gamma_{11}^1 = \frac{\psi}{v} \times \frac{d}{dv}\left(\frac{v}{1-v^2/c^2}\right)\frac{dv}{dT'}$$

Now writing the geodesic equation, for our case, we find

$$a^1 = \frac{d^2 x^1}{d\lambda^2} + \Gamma_{11}^1 \frac{dx^1}{d\lambda} \frac{dx^1}{d\lambda} = 0$$

We clearly see that we can take the affine parameter λ the same as proper time T'. So, we have

$$a^1 = \frac{d^2 x^1}{dx^{1^2}} + \Gamma_{11}^1 \frac{dx^1}{dx^1} \frac{dx^1}{dx^1} = 0$$

$$a^1 = 0 + \Gamma_{11}^1 \times 1 = 0$$

$$a^1 = \Gamma_{11}^1 = 0$$

$$a^1 = \Gamma_{11}^1 = 0$$

$$a^1 = \frac{\psi}{v} \times \frac{d}{dv}\left(\frac{v}{1-v^2/c^2}\right)\frac{dv}{dT'} = 0$$

$$a_{physical} = \sqrt{g_{11}} a^k = \sqrt{g_{11}}\left[\frac{d^2 x^s}{d\lambda^2} + \Gamma_{pq}^s \frac{dx^p}{d\lambda} \frac{dx^q}{d\lambda}\right] = 0$$

$$a_{physical} = \sqrt{g_{11}} a^1 = \sqrt{g_{11}}\frac{\psi}{v} \times \frac{d}{dv}\left(\frac{v}{1-v^2/c^2}\right)\frac{dv}{dT'} = 0$$

$$a_{physical} = \sqrt{g_{11}} a^1 = \frac{v}{\psi} \times \frac{\psi}{v} \times \frac{d}{dv}\left(\frac{v}{1-v^2/c^2}\right)\frac{dv}{dT'} = 0$$

$$a_{\text{physical}} = \frac{dV}{dT'}\frac{d}{dv}\left(\frac{V}{1-v^2/c^2}\right) = 0$$

$$a_{\text{physical}} = \frac{dV'}{dT'} = \frac{dv}{dT'}\frac{d}{dv}\left(\frac{v}{1-v^2/c^2}\right) = 0$$

$$a_{\text{physical}} = \frac{dV'}{dT'} = \frac{d}{dv}\left(\frac{v}{1-v^2/c^2}\right)\frac{dV}{dT'} = 0$$

$$a_{\text{physical}} = \frac{dV'}{dT'} = \left(\frac{1+\frac{v^2}{c^2}}{(1-\frac{v^2}{c^2})^2}\right)\frac{dV}{dT'} = 0$$

We see that for inertial motion $\frac{dv}{dT'} = 0$. BUT for gravitational counterpart, we have the equation which we have already derived. Yes, you are right, and now look down:

$$\frac{dV}{dT'} = -\frac{GM}{r'^2}\sin\beta\,'$$

$$\frac{dV'}{dT'} = \left(\frac{1+\frac{v^2}{c^2}}{(1-\frac{v^2}{c^2})^2}\right)\frac{dV}{dT'}$$

Similarly, if we use the same TIDLE for ordinary matter under Tahirian Gravitation i.e., dS but with different time and speed, we can show that:

$$dS = \cancel{V}dt$$

$$\frac{dv}{dt} = -\frac{GM}{\cancel{r}^2}\sin\beta$$

$$\frac{d\cancel{v}}{dt} = \left(\frac{1+2\frac{v^2}{c^2}}{(1-2\frac{v^2}{c^2})^2}\right)\frac{dV}{dt} = 0$$

To find the deflection of a photon when it passes near the sun or any other mass, we need to consider the following observations made by Tahir:

1   The speed of transverse waves in the theory of electromagnetic waves is the speed of the Electromagnetic Gravitational Wave (EGW) represented by s.
2   EGW possesses just the wave nature and doesn't like to carry any particle with them during their propagation. Why! Because they are the carriers of the gravitational information themselves and are exempt from being deflected by another gravitational field of matter. Hence, there is no matter particle associated with electromagnetic gravitational waves. There is no such thing as graviton! Particle of matter is associated with light only and, of course, light is also a speed of information, but its purpose is different than that of the purpose of gravitational waves. The clock associated with the electromagnetic gravitational wave and light waves is the coordinate clock

(t or T) and not the proper time clock (T'). I will not go further down into the properties of light otherwise focus from the transverse waves would wander.

3   We started the Tahirian Thought Experiment with Coordinate times but came across proper time in the middle of the equations. In order, to get rid of the effects of proper time dilation we will play around with a different TIDLE till the very end as will follow.
4   Due to the duality of light, light consists of two things, namely a particle of matter and a transverse electromagnetic wave. Particle of matter associated with the transverse wave of light does not move either transversely like the electric and magnetic wave fluctuations or in the form of a wave pattern moving longitudinally but rather it moves in a straight-line path without fluctuating in the transverse direction like ordinary matter does when we throw it. Light carries a coordinate time clock t (or T), *due to the wave nature of light.* Hence no Proper Time Dilation $T' = 0$; $T'_1 = 0, T'_2 = 0, ... \& dT' = 0$. The deflection of a photon in a gravitational field is due to the material nature of Photon whereas the double deflection of a photon is due to the material particle nature plus the wave nature of light. Coordinate clock on the light wave doesn't undergo proper time dilation like the clock on the electromagnetic gravitational wave. Theory of transverse waves concludes the speed of the Electromagnetic Gravitational Waves (EGW) which possess just the wave nature and doesn't like to carry any particle with them during their propagation.
5   When we measure anything, say the length of any object with the speed of light, *when travelling at the speed of light,* then we utilize the wave nature of light-wave to measure, which we always do even travelling at speeds below that of light. The clock associated with the wave nature of light is the Coordinate clock and NOT the proper time clock. Therefore, the proper time interval vanishes, thereby making the proper length interval **dS'** vanish too with it. Hence, we don't see anything (zero length of objects) in the direction of motion at least. Nonetheless, I still believe the phenomenon of zero length is applicable in all directions. Meaning that we cannot see anything while travelling at the speed of light simply. We will probably see darkness everywhere. Isn't it an irony of fate that the same light that brings information to us, but when something travels at the same speed, can't see anything itself except darkness!
6   Tahirian mathematical calculations are as follows:

$$dS = VdT = V'dT'$$

$$dS = V'dT'$$

$$dS \times (1-v^2/c^2) = VdT' = dS' = 0 \; (\text{photon} - \text{wave } v = c)$$
$$(\text{TIDLE for photon-wave})$$

When V approaches C, then dT' approaches zero. dS doesn't approach zero but rather dS'. The light-wave is still going on

Natural Theory of Relativity, Inertia, Gravitation & Gravitivity

at V=C. Light wave is just a fluctuation, analogous to, *but not congruent to*, a ripple in water, a domino -effect or a wave on a string. Care should be exercised in that the light-wave is like such phenomena and never ever congruent to it! Now keeping in view, the above new interval as below, we can re-differentiate the equation below to obtain:

$$dS' = VdT'$$

$$T' = t + \frac{rv\sin\beta}{c^2}$$

$$\frac{dT'}{dT'} = \frac{dt}{dT'} + \frac{d}{dT'}\left(\frac{rv\sin\beta}{c^2}\right)$$

We obtain:

$$dT' = \frac{dt}{1 - \frac{v^2}{c^2}}$$

$$dt = \left(1 - \frac{v^2}{c^2}\right)^2 dT$$

$$dT = \frac{dt}{\left(1 - \frac{v^2}{c^2}\right)^2}$$

In the above calculations, I have utilized the following facts according to the new TIDLE:

$$\frac{dr}{dT'} = V\sin\beta \quad \& \quad \frac{d\beta}{dT'} = \frac{V\cos\beta}{r}$$

The above TIDLE is the only TIDLE in Tahirian Cosmos that has one of the characteristics that at the velocity of the light-wave it gives zero distance travelled in the direction of propagation of wave. Like I said that the proper time is zero due to the wave nature of light, i.e., fluctuations in electric and magnetic fields transverse to the propagation of the wave. So as per the laws of the transverse wave nothing travels in the longitudinal direction. It might mean that ordinary matter bursts at the speed of light into photons travelling transversally to the direction of propagation of the EM-transverse wave!

So, we have got the tool to work.

$$dT\left(1 - \frac{v^2}{c^2}\right) = dT' = \frac{dt}{\left(1 - \frac{v^2}{c^2}\right)} \quad \text{where} \quad dT > dT' > dt$$

$$dT\left(1 - \frac{v^2}{c^2}\right) = \frac{dt}{\left(1 - \frac{v^2}{c^2}\right)}$$

Multiplying the above equation with v we obtain

$$vdT\left(1 - \frac{v^2}{c^2}\right) = \frac{vdt}{\left(1 - \frac{v^2}{c^2}\right)}$$

$$dS\left(1 - \frac{v^2}{c^2}\right) = \frac{vdt}{\left(1 - \frac{v^2}{c^2}\right)}$$

$$dS' = vdT\left(1 - \frac{v^2}{c^2}\right) = \frac{vdt}{\left(1 - \frac{v^2}{c^2}\right)} = 0 \quad \text{(photon − wave at v = c)}$$

Or we can also write as below:

$$dS' = vdT\left(1 - \frac{v^2}{c^2}\right) = \left\{\frac{v}{1 - \frac{v^2}{c^2}}\right\}dt = 0 \quad \text{(photon − wave at v = c)}$$

$$dS' = vdT\left(1 - \frac{v^2}{c^2}\right) = \cancel{V}dt = 0 \quad \text{(photon − wave at v = c)}$$

Where:

$$\cancel{V} = \frac{v}{1 - \frac{v^2}{c^2}}$$

$$\frac{d\cancel{V}}{dt} = \left(\frac{1 + \frac{v^2}{c^2}}{(1 - \frac{v^2}{c^2})^2}\right)\frac{dV}{dt}$$

Or

$$\frac{d\cancel{V}}{dT} = \left(\frac{1 + \frac{v^2}{c^2}}{(1 - \frac{v^2}{c^2})^2}\right)\frac{dV}{dT}$$

The expressions below are like each other. However, they are not congruent (equal in all respect)! The difference is evident!

$$dS' = \cancel{V}dt \quad \text{Photon/wave}$$

$$dS = V'dT' \quad \text{ordinary matter}$$

Using the above value, we can say that the momentum of the photon under gravitation is given by the formula:

$$p = m_L\cancel{V}$$

Where m_L is the longitudinal mass of the photon in light.

It is to be understood that the terms light, light-wave and photon should not be used synonymously. Otherwise, it would create alien black holes in Tahirian Physics.

The rate of change of that momentum w.r.t. T is the force on the photon as below:

$$F = \frac{dp}{dT} = \frac{d}{dT}(m_L\cancel{V}) = m_L\frac{d\cancel{V}}{dT}$$

$$F = m_L\frac{d\cancel{V}}{dT} = m_L\frac{d\cancel{V}}{dt}\frac{dt}{dT} \quad \text{(Chain Rule)}$$

$$F = m_L\frac{d\cancel{V}}{dT} = m_L\frac{d\cancel{V}}{dt}\frac{dt}{dT}$$

$$F = m_L \frac{d\mathbf{V}}{dt}\frac{dt}{dT} = m_L \left(\frac{1 + \frac{v^2}{s^2}}{\{1 - \frac{v^2}{s^2}\}^2} \right) \frac{dV}{dt}\frac{dt}{dT}$$

$$F = m_L \left(\frac{1 + \frac{v^2}{s^2}}{\{1 - \frac{v^2}{s^2}\}^2} \right) \frac{dV}{dt} \left(1 - \frac{v^2}{s^2}\right)^2$$

$$F = m_L (1 + \frac{v^2}{s^2}) \frac{dV}{dt}$$

Where $m_L \frac{dV}{dt} = -\frac{Gm_1 m_L}{r^2}\sin\beta$

$$F_t = +(1 + \frac{v^2}{s^2})\frac{Gm_1 m_L}{r^2}\sin\beta$$

Forces with negative signs are the rectangular components of centrifugal force and with the positive signs are the rectangular components of gravitational force/mechanical force! *Extreme care should be exercised in drawing the directions of the arrows.* The resultant of the gravitational components is centripetal (centre seeking direction) whereas the resultant due to the centrifugal components is centre leaving – opposite to centre seeking!

Putting the value of v = c where c = s

We finally obtain as below:

$$F = +2\frac{Gm_1 m_L}{r^2}\sin\beta$$

Now this is the tangential component of the force of gravity acting on the photon-wave (containing duality of light), the normal component is given by:

$$F_{tangential} = 2\frac{Gm_1 m_L}{r^2}\sin\beta$$

$$F_{normal} = 2\frac{Gm_1 m_L}{r^2}\cos\beta$$

The resultant gravitational force the sun is exerting on the photon and vice versa is:

$$F = \{(2\frac{Gm_1 m_L}{r^2}\sin\beta)^2 + (2\frac{Gm_1 m_L}{r^2}\cos\beta)^2\}^{\frac{1}{2}}$$

$$F = 2\frac{Gm_1 m_L}{r^2}$$

The deflection of the photon is double the gravitational value because the force attracting the photon towards the gravitating mass becomes double due to the duality of light, i.e., the wave and the particle nature of the light.

It can also be believed that dark matter consists of cool and low energy photons which, when given enough energy, trigger observable and non-observable waves of light.

When a photon-wave passes near the sun, it experiences gravitational force which is doubled due to a difference of measured time T, which itself is t, for gravitational information by the light wave clock and the coordinate time t of the EGW itself. What an explosive inter-convergence of t n T!

When a body of matter at rest or in uniform motion in a straight line (in an inertial frame) is accelerated then it keeps on accelerating in deep space as long as there is rate of change of momentum or in other words force being constantly applied to the body. Therefore, as soon as the force is vanished the rate of change is vanished, and the acceleration vanishes and now the same body comes back into an inertial frame but this time it keeps on moving in uniform motion in a straight line with the maximum velocity that it got from previous acceleration. We can conclude that it will never keep on going under acceleration because the force has been removed. Therefore, the body will never achieve greater velocity till we apply greater force. There was absolutely no need to put a cosmic speed limit.

"A body continues in its state of uniformly accelerated motion in a straight line as long as the compelling force is acting otherwise."

Now let us generalize the Tahirian Thought Experiment as went before in Figure 1, as below in figure 21.

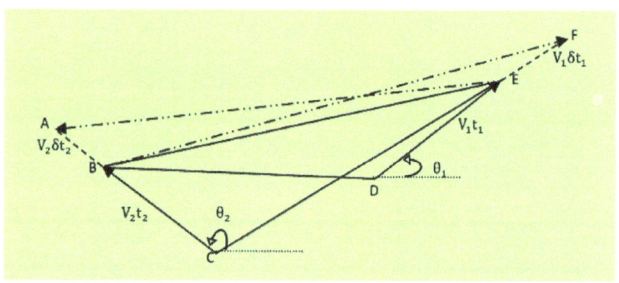

Figure 22. *Relativistic inertial time dilation. Deriving the measured time interval T of an inertial clock, but ticking like a coordinate clock, as a limit with the speed of our information in relativistic motion.*

Let us assume that two objects are moving in two different directions at two different velocities in inertia. One object moves from C to B in time t and the other object moves from D to E in the same amount of time t. The time t is the time of the journeys of both inertial objects shown by the coordinate clocks placed at the ends B and E. I believe that coordinate clocks should better be called the spacetime clocks because they are invisible clocks placed at every point in the fabric of space and don't tick like a clock running in inertia with a constant velocity and ticking slower. I also believe that frames of reference should better be represented by moving objects rather than a fixed system of coordinates representing moving and stationary objects. A moving or stationary object itself constitutes a frame of reference while considering measuring from one frame of reference, which is itself the object, to another frame of reference which in turn is again the object itself. It makes life much easier this way.

If we consider a moving frame approaching a stationary object in inertia, where there is no gravity, a person at rest with respect to the moving frame can say that the object, which is at absolute rest, is approaching him at the same speed but in the opposite direction and hence its clock is ticking slower, whereas its clock is ticking like a coordinate clock. There is absolutely no need for considering moving frames with the help of coordinates because a stationary or a moving object itself constitutes a frame of reference. *So, we can say that "NO TWO" objects moving at different velocities v_1, & v_2 constitute two identical or equivalent or indistinguishable inertial frames of references. Moreover, the laws of physics are the same in all these unidentical, non-equivalent and distinguishable inertial frames of references.* I will give ample proofs of this fact in the coming discussions.

The entire universe is filled up with hypothetical coordinate clocks at all coordinates of the universal grid encompassing the entire cosmos in three dimensions. They measure with infinite speed of information and their time is represented with T.

The equations related to the two frames of references as shown in the above figure are as below:

$$\delta t_1 \neq \delta t_2$$
$$\delta S_1 = v_1 \delta t_1$$
$$\delta S_2 = v_2 \delta t_2$$
$$t_1 = t_2$$
$$dt_1 = dt_2$$
$$dS_1 = v_1 dt_1$$
$$dS_2 = v_2 dt_2$$

$\delta t_1 \neq \delta t_2$ because the objects are moving at two different velocities. The time it takes for the information speed c to bring information from one frame of reference (sender) to the other (receiver) depends upon the speed of the receiver.

Utilizing the previous equation as went before, we have:

$$T = \frac{T'}{\left(1 - \frac{v^2}{c^2}\right)} = \frac{t}{\left(1 - 2\frac{v^2}{c^2}\right)}$$

Now in our current scenario, we are dealing with two objects moving in two different directions with two different velocities. Hence, we need to use subscripts to differentiate one object from the other. So, in line with what has been said we have:

$$T_1 = \frac{T'_1}{\left(1 - \frac{v_1^2}{c^2}\right)} = \frac{t_1}{\left(1 - 2\frac{v_1^2}{c^2}\right)}$$

&

$$T_2 = \frac{T'_2}{\left(1 - \frac{v_2^2}{c^2}\right)} = \frac{t_2}{\left(1 - 2\frac{v_2^2}{c^2}\right)}$$

In the above equations v is a constant because the two bodies are exchanging information in inertia where there is no intervention of gravitation. Hence, we can differentiate the above equations to obtain:

$$dT_1 = \frac{dT'_1}{\left(1 - \frac{v_1^2}{c^2}\right)} = \frac{dt_1}{\left(1 - 2\frac{v_1^2}{c^2}\right)}$$

&

$$dT_2 = \frac{dT'_2}{\left(1 - \frac{v_2^2}{c^2}\right)} = \frac{dt_2}{\left(1 - 2\frac{v_2^2}{c^2}\right)}$$

If we multiply the first equation above by, v_1, we obtain:

$$v_1 dT_1 = \frac{v_1 dT'_1}{\left(1 - \frac{v_1^2}{c^2}\right)} = \frac{v_1 dt_1}{\left(1 - 2\frac{v_1^2}{c^2}\right)}$$

The above equation can also be written as

$$v_1 \left(1 - 2\frac{v_1^2}{c^2}\right) dT_1 = \frac{v_1 \left(1 - 2\frac{v_1^2}{c^2}\right) dT'_1}{\left(1 - \frac{v_1^2}{c^2}\right)} = v_1 dt_1 = dS_1$$

Let us introduce other velocities for further understanding of relativity as v'_1, v''_1 & v'''_1 as below:

$$v''_1 dT_1 = v'''_1 dT'_1 = v_1 dt_1$$

$$v'_1 = \frac{v_1}{\left(1 - \frac{v_1^2}{c^2}\right)}$$

$$v''_1 = v_1 \left(1 - 2\frac{v_1^2}{c^2}\right)$$

$$v'''_1 = \frac{v_1 \left(1 - 2\frac{v_1^2}{c^2}\right)}{\left(1 - \frac{v_1^2}{c^2}\right)}$$

v'_1 is the equivalent velocity recorded (i.e., measured with infinite speed of information) by the proper time clock in the object 1 frame *of* the velocity v_1 recorded by coordinate or spacetime clocks fixed in the fabric of space with infinite speed of information.

v''_1 is the velocity measured with c and coordinate or spacetime clock *moving with the object subscripted 1 but*

ticking like a spacetime clock of the same objects coordinate or spacetime velocity v_1 recorded by coordinate or spacetime clocks fixed in the fabric of space with infinite speed of information.

v_1''' is the velocity measured with c and proper time clock, ticking slower in inertial motion, in the moving frame of object subscripted 1 of the same objects coordinate velocity v_1 recorded by coordinate or spacetime clocks fixed in the fabric of space with infinite speed of information.

When we mean infinite speed of information, we simply mean that the factor of measurement from one frame to the other i.e.,

$$\left(1 - 2\frac{v_1^2}{c^2}\right)$$

approaches 1. We must take into consideration this aspect of the infinite speed of information otherwise we would go into unending cyclic substitutions which won't make a big difference because relativity or gravitivity is a negligibly small part of gravitation. It is inherently very important to conceive ideas with the infinite speed of information.

Now let us multiply the above equation with v_2 as below and see what it returns this time.

$$v_2 dT_1 = \frac{v_2 dT'_1}{\left(1 - \frac{v_1^2}{c^2}\right)} = \frac{v_2 dt_1}{\left(1 - 2\frac{v_1^2}{c^2}\right)}$$

$$v_2 dT_1 = \frac{v_2 dT'_1}{\left(1 - \frac{v_1^2}{c^2}\right)} = \frac{v_2 dt_1}{\left(1 - 2\frac{v_1^2}{c^2}\right)}$$

$$v_2\left(1 - 2\frac{v_1^2}{c^2}\right)dT_1 = \frac{v_2\left(1 - 2\frac{v_1^2}{c^2}\right)dT'_1}{\left(1 - \frac{v_1^2}{c^2}\right)} = v_2 dt_1 = v_2 dt_2 = dS_2$$

$$v_2' = \frac{v_2}{\left(1 - \frac{v_1^2}{c^2}\right)}$$

$$v_2'' = v_2\left(1 - 2\frac{v_1^2}{c^2}\right)$$

$$v_2''' = \frac{v_2\left(1 - 2\frac{v_1^2}{c^2}\right)}{\left(1 - \frac{v_1^2}{c^2}\right)}$$

v_2' is the equivalent velocity recorded (i.e., measured with infinite speed of information) by the proper time clock in

the object 1 frame *of* the velocity v_2 recorded by coordinate or spacetime clocks fixed in the fabric of space with infinite speed of information.

v_2'' is the velocity measured with c and coordinate or spacetime clock, moving with the object subscripted 1 but ticking like a spacetime clock, *of* the object 2 coordinate or spacetime velocity v_2 recorded by coordinate or spacetime clocks fixed in the fabric of space with infinite speed of information.

v_2''' is the velocity measured with c and proper time clock, ticking slower in inertial motion, in the moving frame of object subscripted 1 of the object 2 coordinate velocity v_2 recorded by coordinate or spacetime clocks fixed in the fabric of space with infinite speed of information.

The above discussion is for particles travelling much slower than the speed of light. For particles travelling at the speed of light, i.e., a photon, we have a different set of equation as below:

$$T = \frac{T'}{\left(1 - \frac{v^2}{c^2}\right)} = \frac{t}{\left(1 - \frac{v^2}{c^2}\right)^2}$$

After differentiation we have

$$dT = \frac{dT'}{\left(1 - \frac{v^2}{c^2}\right)} = \frac{dt}{\left(1 - \frac{v^2}{c^2}\right)^2}$$

After giving subscript, we have

$$dT_1 = \frac{dT'_1}{\left(1 - \frac{v_1^2}{c^2}\right)} = \frac{dt_1}{\left(1 - \frac{v_1^2}{c^2}\right)^2}$$

&

$$dT_2 = \frac{dT'_2}{\left(1 - \frac{v_2^2}{c^2}\right)} = \frac{dt_2}{\left(1 - \frac{v_2^2}{c^2}\right)^2}$$

If we multiply the first equation above by c, we obtain:

$$cdT_1 = \frac{cdT'_1}{\left(1 - \frac{v_1^2}{c^2}\right)} = \frac{cdt_1}{\left(1 - \frac{v_1^2}{c^2}\right)^2}$$

The above equation can also be written as

$$c\left(1 - \frac{v_1^2}{c^2}\right)^2 dT_1 = c\left(1 - \frac{v_1^2}{c^2}\right)dT_1' = cdt_1 = dS_1$$

$$c'' = c\left(1 - \frac{v_1^2}{c^2}\right)^2$$

$$c''' = c\left(1 - \frac{v_1^2}{c^2}\right)$$

c'' is the velocity measured with c and coordinate or spacetime clock, moving with the object subscripted 1 but ticking like a spacetime clock, *of* the object 2 (photon here in this case) coordinate or spacetime velocity c recorded by coordinate or spacetime clocks fixed in the fabric of space with infinite speed of information.

c''' is the velocity measured with c and proper time clock, ticking slower in inertial motion, in the moving frame of object subscripted 1 of the object 2 (photon here in this case) coordinate velocity c recorded by coordinate or spacetime clocks fixed in the fabric of space with infinite speed of information.

If we insert c in place of v_1 in the above equations, we see that:

$$c'' = c\left(1 - \frac{v_1^2}{c^2}\right)^2 = 0$$

$$c''' = c\left(1 - \frac{v_1^2}{c^2}\right) = 0$$

There is absolutely no exchange of information between two photons. It seems to be an irony of fate that even though photons bring information but themselves are left uninformed of each other.

Now let us prove that the laws of physics are the same in all distinguishable frames of reference. We will take the example of the velocity as discussed above.

$$v_1' = \frac{v_1}{\left(1 - \frac{v_1^2}{c^2}\right)}$$

The momentum of the object is given by:

$$p_1' = m_1 v_1'$$

Kinetic energy is given by the formula:

$$K.E = \int_0^v FdScos\theta$$

$$K.E = \int_0^v FdS$$

Where θ is zero.

$$K.E = \int_0^v \frac{dP_1'}{dT_1'} dS_1'$$

$$K.E = \int_0^v dP_1' \frac{dS_1'}{dT_1'}$$

$$K.E = \int_0^v \frac{dS_1'}{dT_1'} dP_1'$$

$$K.E = \int_0^v V_1' dP_1'$$

$$K.E = \int_0^v V_1' dP_1' = \left| V_1' P_1' - \int P_1' dV_1' \right|_0^v$$

$$K.E = \left| V_1' m_1 V_1' - \int m_1 V_1' dV_1' \right|_0^v$$

$$K.E = \frac{1}{2} m_1 V_1^2$$

Or

$$K.E = \frac{m_1 V_1^2}{2\left(1 - \frac{v_1^2}{c^2}\right)^2}$$

Whichever way we calculate the K.E it turns out to be the same in all distinguishable inertial frames moving with $V_1', V_2', V_3' \ldots \ldots V_n'$.

$$K.E = \frac{1}{2} m_1 V_1'^2$$

$$K.E = \frac{1}{2} m_2 V_2'^2$$

.

.

.

$$K.E = \frac{1}{2} m_n V_n'^2$$

The laws of physics are the same in all distinguishable inertial frames of references moving at different inertial velocities.

The speed of light is the same only in the fabric of spacetime i.e., recorded the same by the spacetime clocks or coordinate clocks with infinite speed of information. This postulate is the "CRUX" of my theory.

$$c'' = c\left(1 - \frac{v_1^2}{c^2}\right)^2$$

$$c''' = c\left(1 - \frac{v_1^2}{c^2}\right)$$

Or more precisely

$$c'' = c\left(1 - \frac{v_1^2}{s^2}\right)^2$$

$$c''' = c\left(1 - \frac{v_1^2}{s^2}\right)$$

$$p''' = m_1 c'''$$

The speed of light is "NOT" the same for all inertial observers in distinguishable inertial frames. It is measured differently by both proper time and coordinate time clocks with finite speed of information c.

Analogously, we can also say that the speed of gravitational information is "NOT" the same for all inertial observers in distinguishable inertial (instantaneously gravitational) frames. It is measured differently by both proper time and coordinate time clocks with finite speed of gravitational information s as below.

$$s'' = s\left(1 - \frac{v_1^2}{s^2}\right)^2$$

$$s''' = s\left(1 - \frac{v_1^2}{s^2}\right)$$

Inertial frames are easily distinguishable from each other. What distinguishes them is their different inertial velocities with which they are moving.

In the same fashion as went before, we can also show that the kinetic energy of a photon is given by,

$$K.E = \frac{1}{2}m_1 c^2$$

$$K.E = \frac{1}{2}m_1 c_1'''^2$$

.

.

.

$$K.E = \frac{1}{2}m_1 c_n'''^2$$

Where:

$$c_n''' = c\left(1 - \frac{v_n^2}{c^2}\right) \quad \& \quad c = s$$

The photon accelerates right after its existence with a colossal acceleration in the tiniest fraction of a second to the speed of light.

So, the quantum of light is given, as per Natural Theory of Relativity & Gravitivity, as below:

$$K.E = \frac{1}{2}m_1 c^2$$

$$= \frac{1}{2} * 1.295 \times 10^{-32} \text{ kg} * (299792458\tfrac{m}{s})^2$$
$$\sim 5.819 * 10^{-16} \text{ J}$$

We can visualize this by slightly hitting a billiard ball by another billiard ball in a straight line. At low speeds we can see the billiard ball accelerating but hitting the ball at very fast speed, it's very difficult to visualize the almost instantaneous acceleration of the ball. Similarly, we can't visualize and perceive photon balls accelerating at colossal acceleration almost instantaneously right after they get the required quantum. I believe that the acceleration of the billiard ball is approximately as smaller than that of the acceleration of the photon as is the mass of photon is smaller than that of the mass of the billiard ball.

Ibne Sina-Ibn-ur-Rushd- noton Commentary

The applied force is directly proportional to the rate of change of momentum and acts in the direction in which the change acts.

Laws of motion

1a. A body continues in its state of rest or of uniform motion in a straight line unless compelled by some external force to act otherwise.

1b. A body continues in its state of uniformly accelerated motion in a straight line if the compelling force is acting otherwise.
   or
   A body continues in its state of accelerated motion in a straight or curved line if the compelling force is acting otherwise.

1c. A body continues in its state of uniform (constant angular speed) rotational/spinning motion about a fixed axis unless compelled by some internal or external force to act otherwise.

2 To every action there is an equal and opposite

reaction.

3 The applied force is directly proportional to the rate of change of momentum and acts in the direction in which the change acts.

7. Tests of NTRIGG

7.1. Bending of Ray of Light

Using Tahirian equation of closed planetary orbits, we have:

$$r = \frac{h^2/Gm_1}{1 - \varepsilon\sin\theta}$$

$$r = \frac{h^2/Gm_1}{1 + \varepsilon\sin\theta}$$

$$r = \frac{h^2/Gm_1}{1 - \varepsilon\cos\theta}$$

$$r = \frac{h^2/Gm_1}{1 + \varepsilon\cos\theta}$$

All four forms are equally good to represent the elliptical trajectories of static non-gravitivistic orbits depending upon the orientation we want.

So, using the second one, we have for $\theta = +90^o$

$$r = \frac{h^2/Gm_1}{1 + \varepsilon}$$

But for photon, we have

$$r = \frac{h^2/2Gm_1}{1 + \varepsilon}$$

When $\theta = +90^\circ$ then $\beta = 0^\circ$. Setting $r = R$ and $v = c$ we obtain for photon:

$$\varepsilon = \frac{Rc^2}{2GM} - 1 \cong \frac{Rc^2}{2GM} \qquad \text{Where} \qquad \varepsilon \ggg 1$$

Inserting this value of eccentricity ε from above equation in the other hyperbolic equation, as will come next in article 9, we have:

$$\delta = \frac{2}{\varepsilon} = \frac{4GM}{Rc^2} \quad \text{radian}$$

$$\delta = \frac{2}{\varepsilon} = \frac{4GM}{Rc^2} \times \frac{360}{2\pi} \times 3600 \text{ arcsec}$$

The following data plugged in the above equation gives the bending of ray of light when it passes near the sun to be:

$$\delta = \frac{4GM}{Rc^2} \times \frac{360}{2\pi} \times 3600 = 1.75 \text{ arcsec}$$

Where:

$$G = 6.67 \times 10^{-11} \text{ N.m}^2/\text{kg}^2$$

$$M = 1.99 \times 10^{30} \text{ kg}$$

$$R = 695,500,000 \text{ m}$$

$$c = 299,792,458 \text{ m/s}$$

According to NR when the photon passes near the sun the relativistic gravitational acceleration of the photon increases to double which in turn increases the gravitivistic gravitational force between them. Therefore, the gravitivistic force of gravitation becomes double in the case of the Sun and photon interaction due to gravitational motion and coordinate time light-wave clock which eventually deflects the photon by double the amount as given by the non-gravitivistic gravitation.

7.2. Precession of the Perihelion of Mercury – Gravitivistic – See Article 5

7.3. Gravitivistic Precession of the Perihelion of Mercury due to the tugs of other planets.

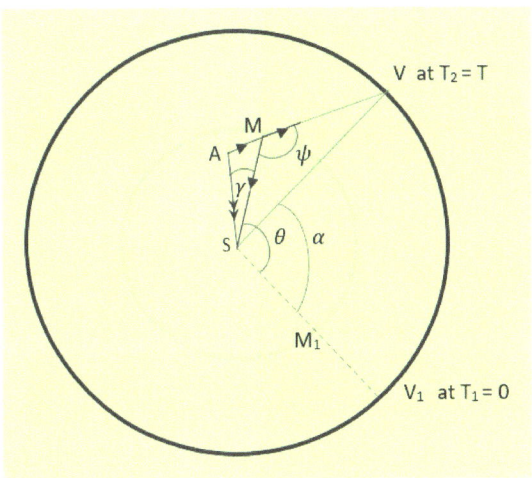

Figure 23. Gravitivistic Precession of the Perihelion of Mercury due to the tugs of other planets.

In this Article, I will find and calculate the gravitivistic precession of the perihelion of Mercury due to the tugs of other planets. In the above figure M_1 represents the initial position of Mercury at time $T_1=0$ and V_1 the initial position of Venus at time $T_1=0$. I have considered circular orbits for simplicity. At time $T_1=0$ Mercury and Venus lie on the same straight line.

Tahir starts the stopwatch and the two starts moving in the counterclockwise direction. After some time at $T_2 = T$ Mercury goes at M and Venus goes at V as shown in the figure above. *A is the perturbed position of Mercury due to the perturbing effect of Venus.*

There are two different triangles in the above figure. One is the force triangle Δ **SMA** and the other is the distance triangle Δ **SMV**.

Angle $M\widehat{S}V = \theta - \alpha$ and angle $S\widehat{M}V = \psi$

To break the force between Venus and Mercury into its rectangular components, let me draw a line (not shown) from point A perpendicular to the line SM and cutting the line at point D not shown in the figure. So, $\overline{AD}$ is perpendicular to $\overline{SM}$. Let me consider the force triangle ΔSMA:

$$tan\gamma = \frac{\kappa_{mv}\dfrac{GM_vM_m}{s^2}sin\,(\pi - \psi)}{\kappa_{ms}\dfrac{GM_sM_m}{r^2} - \kappa_{mv}\dfrac{GM_vM_m}{s^2}cos(\pi - \psi)}$$

$$tan\gamma = \frac{\kappa_{mv}\dfrac{M_v}{s^2}sin\,\psi}{\kappa_{ms}\dfrac{M_s}{r^2} + \kappa_{mv}\dfrac{M_v}{s^2}cos\psi}$$

$$tan\gamma = \frac{\kappa_{mv}M_v sin\,\psi}{\kappa_{ms}\dfrac{M_s}{r^2}s^2 + \kappa_{mv}M_v cos\psi}$$

Where distances

$$\overline{SM} = r; \quad \overline{MV} = s \quad \& \quad \overline{SV} = R$$

From the distance triangle Δ **SMV**, we obtain three relations. Two from the Al-Kashi Theorem and the third from the Naseeruddin Al-Tusi Theorem as below:

$$s^2 = r^2 + R^2 - 2rRcos(\theta - \alpha) \qquad 1$$

$$s^2 = r^2 + R^2 - 2rRcos(\omega_m - \omega_v)T$$

$$s = \sqrt{r^2 + R^2 - 2rRcos(\omega_m - \omega_v)T} = f_1(T)$$

$$R^2 = r^2 + s^2 - 2rscos\psi \qquad 2$$

$$sin\psi = \frac{Rsin(\theta - \alpha)}{s} \qquad 3$$

Differentiating the Alkashi theorem both sides w.r.t the variables s & T using differentials, we obtain

$$d(s^2) = d(r^2 + R^2 - 2rRcos(\theta - \alpha))$$

$$d(s^2) = d(r^2 + R^2 - 2rRcos(\omega_m - \omega_v)T)$$

Where $\quad \theta = \omega_m T \,\&\, \alpha = \omega_v T$

$$sds = rRsin(\omega_m - \omega_v)T(\omega_m - \omega_v)dT$$

$$\frac{ds}{dT} = \frac{rRsin(\omega_m - \omega_v)T(\omega_m - \omega_v)}{\sqrt{r^2 + R^2 - 2rRcos(\omega_m - \omega_v)T}} = f_2(T)$$

$$Rsin(\omega_m - \omega_v)T = Rsin(\theta - \alpha) = \frac{s}{r(\omega_m - \omega_v)}\frac{ds}{dT}$$

$$sin\psi = \frac{Rsin(\theta - \alpha)}{s} = \frac{1}{r(\omega_m - \omega_{v)})}\frac{ds}{dT}$$

$$sin\psi = \frac{1}{r(\omega_m - \omega_{v)})}\frac{ds}{dT}$$

$$cos\psi = \frac{s^2 + r^2 - R^2}{2rs}$$

$$tan\gamma = \frac{\kappa_{mv}M_v sin\,\psi}{\kappa_{ms}\dfrac{M_s}{r^2}s^2 + \kappa_{mv}M_v cos\psi}$$

$$tan\gamma = \frac{\kappa_{mv}M_v\dfrac{1}{r(\omega_m - \omega_{v)})}\dfrac{ds}{dT}}{\kappa_{ms}\dfrac{M_s}{r^2}s^2 + \kappa_{mv}M_v\dfrac{s^2 + r^2 - R^2}{2rs}} = f(T)$$

$$tan\gamma\, dT = \frac{\kappa_{mv}M_v\dfrac{ds}{r(\omega_m - \omega_{v)})}}{\dfrac{M_s}{r^2}s^2 + \kappa_{mv}M_v\dfrac{s^2 + r^2 - R^2}{2rs}}$$

$$tan\gamma\, dT = \frac{\kappa_{mv}M_v\dfrac{ds}{r(\omega_m - \omega_{v)})}}{\kappa_{ms}\dfrac{M_s}{r^2}\left(s^2 + \dfrac{\kappa_{mv}M_v r}{\kappa_{ms}M_s}\dfrac{(s^2 + r^2 - R^2)}{2s}\right)}$$

$$tan\gamma\, dT = \frac{\kappa_{mv}M_v r}{\kappa_{ms}M_s}\frac{1}{(\omega_m - \omega_v)}\frac{ds}{\left(s^2 + \dfrac{\kappa_{mv}M_v r}{\kappa_{ms}M_s}\dfrac{(s^2 + r^2 - R^2)}{2s}\right)}$$

Where:

$$\kappa_{mv} = \left[1 + 3\frac{u_{mv}^2}{s^2}\right]$$

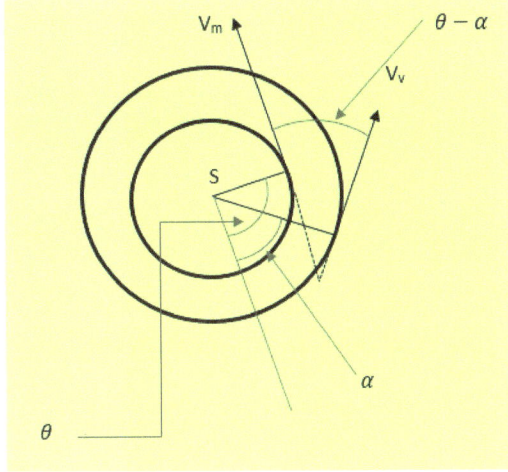

Figure 24. *Angle between the velocities of Mercury and Venus ($\theta - \alpha$).*

$$u_{mv}^2 = v_m^2 + v_v^2 - 2v_m v_v \cos[\pi - [\theta - \alpha]]$$

$$u_{mv}^2 = \omega_m^2 r^2 + \omega_v^2 R^2 + 2v_m v_v \cos(\omega_m - \omega_v)T$$

$$\kappa_{ms} = \left[1 + 3\frac{u_{ms}^2}{s^2}\right]$$

$$u_{ms} = v_m + v_s$$

$$\kappa_{ms} = \left[1 + 3\frac{u_{ms}^2}{s^2}\right]$$

$$u_{ms} = V_m + V_s$$

Where:

$$V_s = \frac{M_m V_m}{M_s}$$

(Velocity of the Sun in Mercury-Sun Binary System)

Tahir Gravitivistic Terms are not considered due to their values being negligible. But for the balancing of Tahirian Gravitational Mutual Force between two particles of matter they are necessary to be discussed.

$$tan\gamma dT = \frac{M_v r}{M_s}\frac{1}{(\omega_m - \omega_v)}\frac{ds}{s^2(1 + \frac{M_v r}{M_s}\frac{(s^2 + r^2 - R^2)}{2s^3})}$$

$$tan\gamma dT = \frac{M_v}{M_s}r\frac{1}{(\omega_m - \omega_v)}\frac{ds}{s^2(1 + \frac{1}{2}\frac{M_v r}{M_s s} + \frac{1}{2}\frac{M_v r}{M_s s^3}(r^2 - R^2))}$$

$$tan\gamma dT = \frac{M_v}{M_s}r\frac{1}{(\omega_m - \omega_v)}\frac{ds}{s^2(1 + \frac{1}{2}\frac{M_v r}{M_s s} + \frac{1}{2}\frac{M_v r^3}{M_s s^3} - \frac{1}{2}\frac{M_v rR^2}{M_s s^3})}$$

Due to the mass of Venus being very small as compared to the mass of the Sun, I can apply the Omar Khayyam's Binomial series expansion as below:

$$tan\gamma dT = \frac{M_v}{M_s}r\frac{1}{(\omega_m - \omega_v)}\frac{\left[1 + \frac{1}{2}\frac{M_v r}{M_s s} + \frac{1}{2}\frac{M_v r^3}{M_s s^3} - \frac{1}{2}\frac{M_v rR^2}{M_s s^3}\right]^{-1}ds}{s^2}$$

$$tan\gamma dT = \frac{M_v}{M_s}r\frac{1}{(\omega_m - \omega_v)}\frac{\left[1 - \frac{1}{2}\frac{M_v r}{M_s s} - \frac{1}{2}\frac{M_v r^3}{M_s s^3} + \frac{1}{2}\frac{M_v rR^2}{M_s s^3}\right]ds}{s^2}$$

$$tan\gamma dT = \frac{M_v}{M_s}r\frac{1}{(\omega_m - \omega_v)}\frac{\left[1 - \frac{1}{2}\frac{M_v r}{M_s s} - \frac{1}{2}\frac{M_v r^3}{M_s s^3} + \frac{1}{2}\frac{M_v rR^2}{M_s s^3}\right]ds}{s^2}$$

$$tan\gamma dT = \frac{M_v}{M_s}r\frac{1}{(\omega_m - \omega_v)}\left[\frac{1}{s^2} - \frac{1}{2}\frac{M_v r}{M_s s^3} - \frac{1}{2}\frac{M_v r^3}{M_s s^5} + \frac{1}{2}\frac{M_v rR^2}{M_s s^5}\right]ds$$

Integrating both sides of the equation, we obtain as below:

$$\int_0^{2\pi/\omega_m} tan\gamma dT = \frac{M_v}{M_s}r\frac{1}{(\omega_m - \omega_v)}\int_0^{2\pi/\omega_m}\left[\frac{1}{s^2} - \frac{1}{2}\frac{M_v r}{M_s s^3} - \frac{1}{2}\frac{M_v r^3}{M_s s^5} + \frac{1}{2}\frac{M_v rR^2}{M_s s^5}\right]ds$$

I have integrated over one cycle time of the two planets.

$$\int_0^{2\pi/\omega_m} tan\gamma dT = \int_0^{2\pi/\omega_m} f(T)dT$$

From Mean-Value theorem

$$\int_a^b f(T)dT = \overline{f(T)}b - a)$$

$$\int_a^b f(T)dT = f(c)b - a)$$

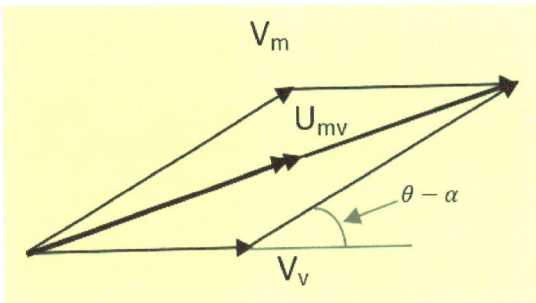

Figure 25. *Resultant velocity of the Mercury-Venus system.*

$f(c)$ or $\overline{f(T)}$ is the mean value of all the possible values of the function $f(T)$ between a and b.

$$\overline{f(T)} = \lim_{n\to\infty}\frac{f(T_1) + f(T_2) + f(T_3) + \cdots\ f(T_n)}{n}$$

$$\overline{f(T)} = \lim_{n\to\infty}\frac{\sum_1^n f(T_n)}{n}$$

$$\overline{tan\gamma} = \lim_{n\to\infty}\frac{tan\gamma_1 + tan\gamma_2 + tan\gamma_3 \ldots\ldots\ldots tan\gamma_n}{n}$$

$$\overline{tan\gamma} \times \frac{2\pi}{\omega_m} = \frac{M_v}{M_s}r\frac{1}{(\omega_m - \omega_v)}\int_0^{2\pi/\omega_m}\left[\frac{1}{s^2} - \frac{1}{2}\frac{M_v}{M_s}\frac{r}{s^3} - \frac{1}{2}\frac{M_v}{M_s}\frac{r^3}{s^5} + \frac{1}{2}\frac{M_v}{M_s}\frac{rR^2}{s^5}\right]ds$$

$$2\pi\overline{tan\gamma} = 2\pi tan\gamma_c = \frac{M_v}{M_s}r\frac{1}{\left(1 - \frac{\omega_v}{\omega_m}\right)}\int_0^{2\pi/\omega_m}\left[\frac{1}{s^2} - \frac{1}{2}\frac{M_v}{M_s}\frac{r}{s^3} - \frac{1}{2}\frac{M_v}{M_s}\frac{r^3}{s^5}\right.$$
$$\left. + \frac{1}{2}\frac{M_v}{M_s}\frac{rR^2}{s^5}\right]ds$$

$$2\pi tan\gamma_c$$
$$= \frac{M_v}{M_s}r\frac{1}{\left(1 - \frac{\omega_v}{\omega_m}\right)}\left[\left|-\frac{1}{s}\right|_0^{2\pi/\omega_m} - \frac{1}{2}\frac{M_v}{M_s}r\left|-\frac{1}{2s^2}\right|_0^{2\pi/\omega_m}\right.$$
$$\left. - \frac{1}{2}\frac{M_v}{M_s}r^3\left|-\frac{1}{4s^4}\right|_0^{2\pi/\omega_m} + \frac{1}{2}\frac{M_v}{M_s}rR^2\left|-\frac{1}{4s^4}\right|_0^{2\pi/\omega_m}\right]$$

γ_c is the angle whose tangent is the mean of the tangents of all the angles γ.

We can truncate the above to the first term only because the second, third and fourth terms are negligible as compared to the first term.

$$2\pi tan\gamma_c = \frac{M_v}{M_s}r\frac{1}{\left(1 - \frac{T_m}{T_v}\right)}\left|-\frac{1}{s}\right|_0^{\frac{2\pi}{\omega_m}} \text{ rad/revolution}$$

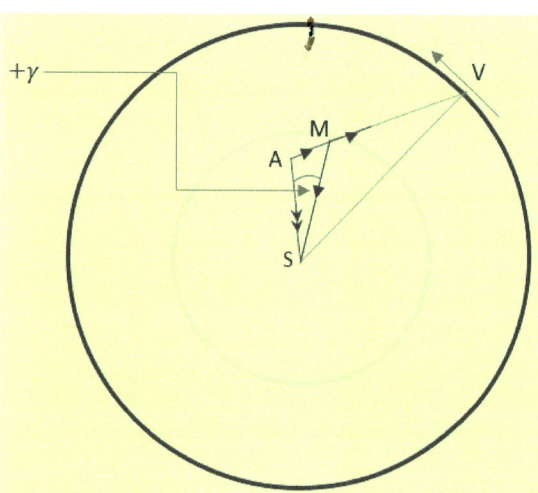

Figure 26. *Mercury leading Venus in the direction of motion shown-counterclockwise. Mercury precessing in the direction of motion. γ is positive.*

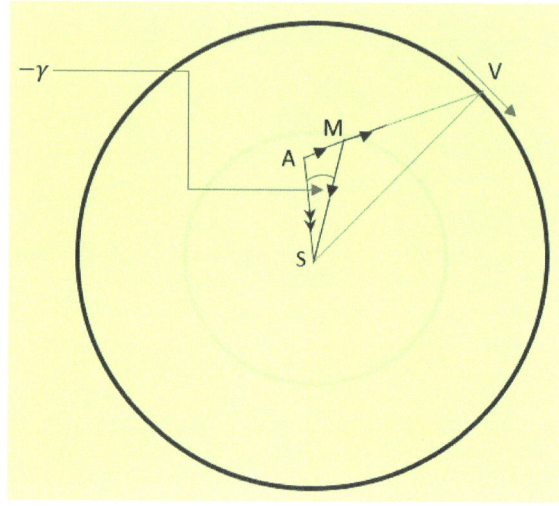

Figure 27. *Venus leading Mercury in the direction of motion shown-clockwise. Mercury precessing opposite the direction of motion. γ is negative.*

$$2\pi tan\gamma_c = \frac{M_v}{M_s}r\frac{1}{\left(1 - \frac{T_m}{T_v}\right)}\left|-\frac{1}{s}\right|_0^{\frac{2\pi}{\omega_m}} \times 414.9 * \frac{360}{2\pi} \times 3600$$
$$\textbf{arcsec/century}$$

Considering the above figures 26 & 27, we can break up the motion of the two bodies Mercury and Venus into two different time intervals, viz-a-viz, the time, during one complete cycle of Mercury, when it leads Venus up-to an angle $\boldsymbol{\beta = \pi}$ and the rest of the time, during one complete cycle of Mercury, when Venus leads Mercury as below:

$$T_1 = \frac{\theta}{\omega_m} = \frac{\alpha}{\omega_v}$$

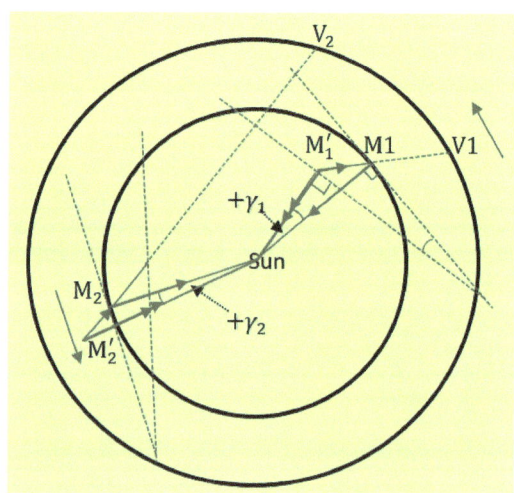

Figure 28. *Mercury leading Venus in the direction of motion shown-counterclockwise for a certain portion of its cycle of 2π radians. Mercury precesses in the direction of motion. γ is positive.*

$$\boldsymbol{\beta = \theta - \alpha}$$

Natural Theory of Relativity, Inertia, Gravitation & Gravitivity

$$\beta = T_1 \omega_m - T_1 \omega_v = \pi$$

$$T_1(\omega_m - \omega_v) = \pi$$

$$T_1 = \frac{\pi}{\omega_m - \omega_v}$$

The above is the time T taken by the two planets to become collinear again with the Sun *but on opposite sides* of the Sun.

$$2\pi tan\gamma_c = \bar{\bar{\gamma}} = \frac{M_v}{M_s} r \frac{1}{\left(1 - \frac{T_m}{T_v}\right)} \left[\left| -\frac{1}{s} \right|_0^{\frac{\pi}{\omega_m - \omega_v}} + \left| -\frac{1}{s} \right|_{\frac{\pi}{\omega_m - \omega_v}}^{\frac{2\pi}{\omega_m}} \right] \times 414.9 * \frac{360}{2\pi} \times 3600$$

So, we see that:

$$+\gamma_1 = \frac{M_v}{M_s} r \frac{1}{\left(1 - \frac{T_m}{T_v}\right)} \left[\left| -\frac{1}{s} \right|_0^{\frac{\pi}{\omega_m - \omega_v}} \right] \times 414.9 * \frac{360}{2\pi} \times 3600 \text{ (Mercury leads)}$$

$$-\gamma_2 = \frac{M_v}{M_s} r \frac{1}{\left(1 - \frac{T_m}{T_v}\right)} \left[\left| -\frac{1}{s} \right|_{\frac{\pi}{\omega_m - \omega_v}}^{\frac{2\pi}{\omega_m}} \right] \times 414.9 * \frac{360}{2\pi} \times 3600 \text{ (Venus leads)}$$

$$2\pi tan\gamma_c = +\gamma_1 - \gamma_2$$

$$\bar{\bar{\gamma}} = +\gamma_1 - \gamma_2$$

Double overbar is shown to distinguish the Al-jebraic value between two quantities of two opposite signs and the average value as discussed earlier. The Al-jebraic value $\bar{\bar{\gamma}}$ is the precession of the Mercury in the forward direction of motion due to the tug of Venus.

	$+\gamma_1$	$-\gamma_2$	$+\bar{\bar{\gamma}}$
Venus	+276.48	-6.59	+269.89
Earth	+119.28	-30.13	+89.15
Mars	+4.37	-3.12	+1.25
Jupiter	+927.54	-922.81	+4.73
Saturn	+80.72	-80.66	+0.06
Uranus	+3.06	-3.05973	+0.00027
Neptune	+1.45	-1.449968	+0.000032
Pluto	+.00011	-0.0001099	+0.000000001
Total	+1412.90	-1047.819	+365.08

Table 2. *Precession of Mercury due to the tugs of other planets in one cycle of Mercury.*

In Table 2 above, we see that the maximum amount of perturbation, in the orbit of Mercury, is caused by Jupiter in both

the direction of motion and against it.

I have calculated the precession of Mercury in one cycle of Mercury, i.e., when it completes one revolution around the Sun. It can be understood that this amount of precession of Mercury is the same in every cycle of the two planets Mercury and Venus irrespective of the position of Venus because the two planets go through the same phase of orbital motion. Let us suppose Venus is ahead of Mercury in orbiting at an instant as per the figure 29 below:

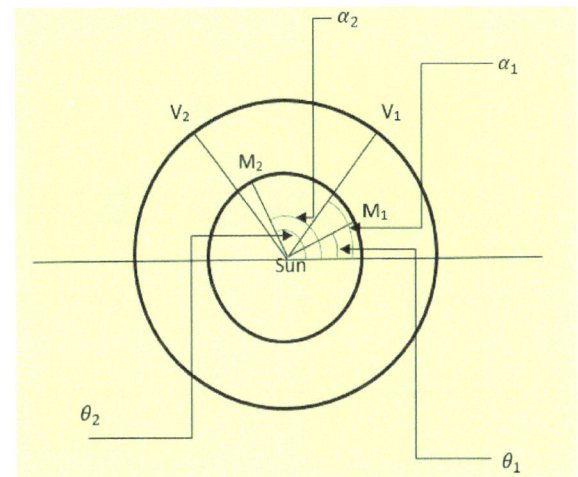

Figure 29. *Equal precession in all cycles of Mercury due to the tug of Venus.*

The angle β between the two planets, at an instant in a cycle, is given by:

$$\beta_1 = \alpha_1 - \theta_1$$

$$\beta_2 = \alpha_2 - \theta_2$$

$$\beta_3 = \alpha_3 - \theta_3$$

.

.

.

$$\beta_k = \alpha_k - \theta_k$$

$$k = 1, 2, 3, \ldots \ldots \infty$$

Differentiating the above equation, we obtain as below:

$$\frac{d\beta_k}{dT} = \frac{d\alpha_k}{dT} - \frac{d\theta_3}{dT}$$

$$\frac{d\beta_k}{dT} = \omega_V - \omega_M = \text{Constant}$$

$$\frac{d\beta_1}{dT} = \frac{d\beta_2}{dT} = \cdots$$

$$\beta_1 \neq \beta_2 \neq \beta_3 \neq \cdots \beta_k$$

The angle is changing but it's rate of change is the same throughout the cycles of Mercury because we are assuming circular orbits. In the precession formula, we come across the rate of change of the angle between the two stars at a particular instant and not the angle itself. TCR is very close to the Centre of the Sun and hence not shown with respect to binary Sun-Mercury system.

Now let us find the precession of Mercury in one cycle of *the two planets* instead of one cycle of Mercury. It is to be understood that Mercury is in binary motion with the Sun whereas it is in perturbed motion with Venus. In perturbed motion, only perturbations happen, and binary laws are held intact. Perturbations are a slight *apparent* modification in the system of planets *in the form of a dipole* whose net precession is zero in one cycle of the two planets. By one cycle of the two planets, I mean when Mercury and Venus become collinear with the Sun again on the same side of the Sun which is the same relative position when their motion was set to begin for our calculations. Perturbations of Mercury due to outer planets is basically apparent/dipole and hence can be called a hoax in astrophysics. Perturbations are there but the net effect is zero in one cycle of the two planets Mercury and Venus in our case of study. Care should be exercised NOT to include the laws pertaining to trinaries, quartinaries, pentanaries etc. systems of stars unless or until they are developed and understood to the Tahirian extent.

	$+\gamma_1$	$-\gamma_2$	$+\bar{\bar{\gamma}}$
Venus	+276.48	-276.48	0.00
Earth	+119.28	-119.28	0.00
Mars	+4.37	-4.37	0.00
Jupiter	+927.54	-927.54	0.00
Saturn	+80.72	-80.72	0.00
Uranus	+3.06	-3.06	0.00
Neptune	+1.45	-1.45	0.00
Pluto	+.00011	-0.00011	0.00
Total	+1412.90	-1412.90	0.00

Table 3. Precession of Mercury due to the tugs of other planets in one cycle of the two planets, i.e., Mercury and Venus.

During one cycle of the two planets Mercury leads Venus and produces precession in the forward direction of motion for the period-of-time T as below

$$0 \leq T \leq \frac{\pi}{\omega_m - \omega_v}$$

Whereas Venus leads Mercury during one cycle of the two planets and produces precession in the backward direction of motion for the period-of-time during the cycle as below

$$\frac{\pi}{\omega_m - \omega_v} \leq T \leq \frac{2\pi}{\omega_m - \omega_v}$$

Let us find the time required to complete one cycle of the two planets. By the time Mercury completes one cycle in $\frac{2\pi}{\omega_m}$ days, Venus has subtended an angle of $2\pi\frac{\omega_v}{\omega_m}$ or $2\pi\frac{T_m}{T_v}$ radians at the Sun. Now let us assume that by the time Mercury catches Venus again in collinearity with the Sun, on the same side of the Sun, Venus has further travelled in its circular orbit and has subtended an angle of α at the Sun. The time taken for Venus to subtend an angle of α is equal to the time taken by Mercury to catch Venus by subtending the angle $2\pi\frac{T_m}{T_v} + \alpha$. Hence, we have as below

$$\frac{2\pi\frac{T_m}{T_v}+\alpha}{\omega_m} = \frac{\alpha}{\omega_v} \qquad (t = \frac{\theta}{\omega})$$

So, the time is:

$$T = \frac{2\pi+2\pi\frac{T_m}{T_v}+\alpha}{\omega_m} \qquad (t = \frac{\theta}{\omega})$$

$$\alpha = 2\pi\frac{T_m}{T_v(\frac{T_v}{T_m} - 1)}$$

$$T = \frac{2\pi}{\omega_m - \omega_v}$$

Where:

$$T_m\omega_m = T_v\omega_v = 2\pi$$

Hence, the time for one complete cycle of the two planets is $\frac{2\pi}{\omega_m-\omega_v}$ as compared to the time for one complete cycle of Mercury which is $\frac{2\pi}{\omega_m}$. I have used T synonymously to represent both time and time-period, so be careful while reading!

$$2\pi tan\gamma_c = \bar{\bar{\gamma}} = \frac{M_v}{M_s}r\frac{1}{(1-\frac{T_m}{T_v})}\left[\left|-\frac{1}{s}\right|_0^{\frac{\pi}{\omega_m-\omega_v}} + \left|-\frac{1}{s}\right|_{\frac{\pi}{\omega_m-\omega_v}}^{\frac{2\pi}{\omega_m-\omega_v}}\right] \times 414.9 * \frac{360}{2\pi} \times 3600 = 0$$

$$\bar{\bar{\gamma}} = +\gamma_1 - \gamma_2 = 0$$

The actual precession in one cycle of the two planets might be a small value due to the orbits being elliptical instead of perfect circles as we have assumed in our calculations. If exact positions of the planets are known at a certain time in the elliptical orbits, the precession can be very complicated to calculate if not impossible.

7.4. Gravitational Shift in the Frequency of Light-Wave and Mass of Photons.

We know that light is basically an electromagnetic wave phenomenon [2, 4] and therefore has wave characteristics along with the particle nature. From Planck's law the energy [1] associated with the wave of light (electromagnetic radiation) is $E = h\nu$.

$$E = h\nu$$

$$dE = hd\nu$$

From the equations gone before we have

$$F = 2\frac{Gm_1m_L}{r^2}$$

To find the work done when a photon goes away from matter in the gravitational field of matter, we have:

$$dW = -Fdr$$

$$dW = -2\frac{Gm_1m_L}{r^2}dr$$

For photon $v = c$ & $c = s$ we have

$$dW = -\frac{2Gm_1m_L}{r^2}dr$$

Comparing the differential for the gravitational and wave energy associated with a photon and light wave we have:

$$dE = dW$$

$$hd\nu = -\frac{2Gm_1m_L}{r^2}dr$$

$$\int_{\nu_1}^{\nu_2} hd\nu = -\int_{r_1}^{r_2} \frac{2Gm_1m_L}{r^2}dr$$

$$h(\nu_2 - \nu_1) = 2Gm_1m_L(\frac{1}{r_2} - \frac{1}{r_1})$$

Gravitational shift in the frequency of light-wave

$$\nu_2 - \nu_1 = \frac{2Gm_1m_L}{h}(\frac{1}{r_2} - \frac{1}{r_1})$$

Some of the following experimental data was taken from a website that I have lost the web name/address of:

$$H = 6.626 \times 10^{-34} \text{ J s}$$

$$G = 6.674 \times 10^{-11} \text{ m}^3\text{kg}^{-1}\text{s}^{-2}$$

$$m_{earth} = 5.972 \times 10^{24} \text{ kg}$$

$$\nu_R - \nu_r = 2.5 \times 10^{-15} \times \nu_r \text{ Hz}$$

$$\nu_r = 3.46 \times 10^{18} \text{ Hz}$$

$$R = 6365 \times 10^3 \text{ m}$$

$$r - R = 22.5 \text{ m}$$

The above data when plugged into the above equation produces a mass of the photon nearly equal to the mass of an electron as below:

$$\sim 1.295 \times 10^{-32} \text{ Kg}$$

Frequency is not associated with the photon, but it is associated with the light wave [5]. Where r_1 & r_2 are distances from the center of the earth. It is remarkable to note that the change in frequency does not depend upon the initial frequency of the light wave.

One may wonder how come the change in frequency formula given in books on modern physics for the light wave is independent of the Planck's constant. The change is an interaction of the wave energy and the gravitation and therefore should have had both the universal constants i.e., G & h and the mass of the photon as well. The fact that the researchers have not been able to find the exact mass of a photon is a different story. The ignoring/ignorance of the mass of photon by the researchers due to its highly negligible value is analogous to ignoring/ignorance of the speed of gravitational information in the static orbits by the aliens.

I believe, if researchers can calculate the small change in the value of the frequency of light wave in the laboratory *accurately* then they would be able to find the exact mass of the photon by using the above formula of Natural relativity. The exact mass of the photon remains to be determined precisely. I have lost the website from where I took the data and calculated the mass of a photon.

8. Orbital Speed

According to Tahirian Cosmology, a body of matter thrown with a velocity tangential to the surface of the earth, the point of projection being at the earth, equal to $v = \sqrt{2\frac{Gm_1}{r}}$, will escape the earth's gravity by making a parabolic path. The range of horizontal tangential velocities greater than $v = \sqrt{\frac{Gm_1}{r}}$ and less than $v = \sqrt{2\frac{Gm_1}{r}}$ will force the object to remain in bounded elliptical orbit around the earth. At a horizontal tangential velocity greater than $v = \sqrt{2\frac{Gm_1}{r}}$ the orbit will become hyperbolic/unbounded. We need an infinite amount of horizontal escape velocity to project a body of matter in an exactly horizontal path along a tangential line at the surface of projection. The governing equation for all types of horizontal tangential velocities at a distance r, where r is greater than R, from the center of the Earth is the following:

$$r = \frac{h^2/Gm_1}{1 + \varepsilon\sin\theta}$$

$$v = \sqrt{(1+\varepsilon)\frac{Gm_1}{r}}$$

Where ε is the eccentricity of the orbit that will be adopted by the non-gravitivistic projectile.

$$v = \sqrt{\frac{Gm_1}{r}} \qquad \varepsilon = 0; \quad \text{circular orbit}$$

$$\sqrt{\frac{Gm_1}{r}} < v < \sqrt{2\frac{Gm_1}{r}} \qquad 0 < \varepsilon < 1; \text{elliptical orbit}$$

$$v = \sqrt{2\frac{Gm_1}{r}} \qquad \varepsilon = 1; \text{parabolic orbit}$$

$$v > \sqrt{2\frac{Gm_1}{r}} \qquad \varepsilon > 1; \text{hyperbolic orbit}$$

9. Hyperbolic Orbit

Let us find the eccentricity in terms of the angle that the two asymptotes (on the right and left of y axis) of the hyperbola make with each other. To do so we need to find the slope of the asymptote at infinity. Let point A be the point on the y axis $A(0, a)$, Point P on the hyperbola as $P(x, y)$ and point $B(x, -a)$ on the horizontal line $y = -a$. Let $\overline{PB}$ be always the perpendicular distance of the point P from the Horizontal line $y = -a$ since it's a condition of eccentricity of conic sections [6]. Point A is fixed, and points P and B vary their positions.

Forming the ratio of two lengths

$$\frac{\overline{AP}}{\overline{PB}} = \varepsilon = \frac{\sqrt{(y-a)^2 + (x-0)^2}}{\sqrt{(x-x)^2 + (y+a)^2}}$$

Rearranging terms we obtain:

$$y^2(1-\varepsilon^2) + x^2 - 2ay(1+\varepsilon^2) + a^2(1-\varepsilon^2) = 0$$

After completing squares, the equation can be written as

$$y(1-\varepsilon^2) = \pm\sqrt{4a^2\varepsilon^2 - x^2(1-\varepsilon^2)} + \frac{a^2(1+\varepsilon^2)^2}{(1-\varepsilon^2)}$$

If we subtract the term $-a(1 + \varepsilon^2)$ from the equation above, we get:

$$y(1-\varepsilon^2) - a(1+\varepsilon^2) = \pm\sqrt{4a^2\varepsilon^2 - x^2(1-\varepsilon^2)} + \frac{a^2(1+\varepsilon^2)^2 - a(1-\varepsilon^4)}{(1-\varepsilon^2)}$$

Differentiating equation w.r.t. x and rearranging terms we have

$$\frac{dy}{dx} = \frac{-x}{y(1-\varepsilon^2) - a(1+\varepsilon^2)}$$

Substituting from one equation into the other equation we have

$$\frac{dy}{dx} = \frac{-x}{\pm\sqrt{4a^2\varepsilon^2 - x^2(1-\varepsilon^2)} + \frac{a^2(1+\varepsilon^2)^2 - a(1-\varepsilon^4)}{(1-\varepsilon^2)}}$$

$$\lim_{x\to\infty}\frac{dy}{dx} = \frac{\pm 1}{\sqrt{(\varepsilon^2 - 1)}}$$

By the equation above the slope of the right asymptote to the hyperbola at infinity is

$$\frac{1}{\sqrt{(\varepsilon^2 - 1)}}$$

If we denote the angle that the asymptote of the hyperbola makes with the x-axis by α then the tangent of this angle α is equal to the slope of the asymptote at infinity

$$\tan\alpha = \frac{1}{\sqrt{(\varepsilon^2 - 1)}} \approx \frac{1}{\varepsilon} \qquad \varepsilon \ggg 1 \text{ (photon)}$$

Now $\tan\alpha \cong \alpha$, where α is very small angle in radian.

$$\alpha \approx \frac{1}{\varepsilon}$$

The angle δ that the asymptotes (on the left and on the right) make with each other is double the angle α i.e. $\delta = 2\alpha$. This is the deviation from the matter-free straight path of Ibne-Sina's first law of motion.

$$\delta \approx \frac{2}{\varepsilon}$$

Hyperbolic deflection angle formula for photon speed

$$\delta \cong \frac{2}{\varepsilon}$$

10. Simple Unification of Gravitation and Electromagnetism

Let us reconsider the Equation in Tahirian Cosmology as below:

$$F = \frac{Gm_1m_2}{r^2}\left(1 + 3\frac{v^2}{s^2}\right) = \frac{Gm_1m_2}{r^2}(1 + 3\varepsilon_0\mu_0 v^2)$$

The above equation contains two constants, namely the constant of gravitation G and the constant of Gravitivity $s = \frac{1}{\sqrt{\varepsilon_0\mu_0}}$. This is a simple proposed unification of gravitation and electromagnetism or a simple correction in the law of gravitation but of paramount importance for the understanding of natural phenomenon of gravitation and its kinematics.

I believe that the part played by electromagnetism is nothing more than a superb natural network of information/communication of gravitational force in relative gravitational binary motion in the form of electromagnetic gravitational waves (EGW). These EGWs are, I believe, like ordinary electromagnetic waves but pole apart different in their function and propagation. The difference can easily be understood if we compare ordinary construction steel with post-tensioned strands of steel wires.

I believe that the information of the generation/reception of EGWs is clearly indicated by the parameter m_1 in the above equation i.e. the central body around which the smaller body is rotating. We are familiar with the fact that accelerating charges generate electromagnetic waves and therefore accelerating masses must also do the same. So, in the scenario of cosmic catastrophe of the sudden vaporization of the sun, the probability of the earth to permanently keep going on in outer space by making a tangential path to its orbit about 8 minutes after the vanishing of the sun and the coming of the last EGW wavefront to the earth is very low or non-existent. Because as soon as the sun supposedly vanishes, the parameter m_1 vanishes in equation 50, but much before the lapse of about 8 minutes, instantaneously another m_1' will be ready to take the position in equation 50 above and will reset the generation of EGW's as per routine of the superb natural network of information of relativistic gravitational force. That m_1' will now be the central mass for the earth to rotate or move, about which the sun was once a rotating or moving mass. The speed of information of the EGW will be the same as s, as before, but this time the information conveyed to the earth by the EGW of m_1' will be different than what it was prior to the supposed catastrophe. Similarly, for the rest of the planets of the once solar system, a brilliantly new system, as delicate as the previous solar system, will be developed automatically. This is the surprising backup that, I believe, nature has in store for the so called, cosmic catastrophe. Conjecturing in line with what I have just said above, we can conclude that our cosmos, and not just our solar system, is under the sway of stable equilibrium even under extreme cases of cosmic catastrophes.

Moreover, the earth and the sun are not just sending the gravitational information, they are also receiving gravitational information because earth is bound to follow the sun and the sun is bound to move around the earth under binary stars system with respect to the TCR.

11. Conclusion

The speed given by recreational calculations of magnetoelectrics pvt ltd by maglel is the speed of gravity with which all matter communicates with each other. The speed of light is the same as the speed of gravity, but light has dual nature whereas speed of gravity is wave nature only. Gravitational wave is not deflected by any gravity but photon – material nature of light, is deflected by gravity.

References

[0]   Science in the medieval Islamic world - Wikipedia

[1]   Tensors, Contravariant and Covariant.
      www.mathpages.com/rr/s5-02/5-02.htm.

[2]   Jackson, Tom, Physics, An Illustrated History of the
      Foundations of Science. Ponderables 100 Breakthroughs
      That changed History. Who did What When.

[3]   Gradshteyn, I.S. & Ryzhik, I.M., The table of integrals,
      series, and products. Seventh edition.

[4]   Spiegel, M.R., Vector Analysis and an Introduction to Tensor
      Analysis. Published by Schaum Outline Series (1980-08-01).

[5]   Beiser, A, Concepts of Modern Physics.

[6]   Dakin, R. I. Porter, Elementary Analysis (1971).